Mapping the Diversity of Nature

Mapping the Diversity of Nature

Edited by

Ronald I. Miller

CHAPMAN & HALL

London · Glasgow · Weinheim · New York · Tokyo · Melbourne · Madras

Published by Chapman & Hall, 2–6 Boundary Row, London SE1 8HN, UK

Chapman & Hall, 2–6 Boundary Row, London SE1 8HN, UK

Blackie Academic & Professional, Wester Cleddens Road, Bishopbriggs, Glasgow G64 2NZ, UK

Chapman & Hall GmbH, Pappelallee 3, 69469 Weinheim, Germany

Chapman & Hall USA, One Penn Plaza, 41st Floor, New York NY 10119, USA

Chapman & Hall Japan, ITP Japan, Kyowa Building, 3F, 2–2–1 Hirakawacho, Chiyoda-ku, Tokyo 102, Japan

Chapman & Hall Australia, Thomas Nelson Australia, 102 Dodds Street, South Melbourne, Victoria 3205, Australia

Chapman & Hall India, R. Seshadri, 32 Second Main Road, CIT East, Madras 600 035, India

First edition 1994

© 1994 Chapman & Hall

Typeset in 10½/12½ pt Sabon by Photoprint, Torquay
Printed in Great Britain by the Alden Press, Oxford

ISBN 0 412 45510 2

A catalogue record for this book is available from the British Library

∞ Printed on permanent acid-free text paper, manufactured in accordance with ANSI/NISO Z39.48–1992 and ANSI/NISO Z39.48–1984 (Permanence of Paper).

Dedication

I would like to dedicate this book to the upcoming generations of people whom I hope will appreciate and care for all the gifts of nature. And I would especially like to dedicate this book to my own special emissary in the next generation, to my main man Max.

Contents

Contents

Contents

Contents

Contributors

Jennifer H. Allen
Environmental Sciences and Public Policy
    Program,
George Mason University,
Fairfax, Virginia, USA

John R. Busby
Environmental Resources Information
    Network,
GPO Box 636,
Canberra ACT 2601,
Australia

Bart R. Butterfield
Idaho Fish & Game,
600 South Walnut,
PO Box 25,
Boise, Idaho 83707–0025, USA

Arthur D. Chapman
Australian Biological Resources Study,
Environmental Resources Information
    Network,
GPO Box 636,
Canberra ACT 2601, Australia

Michael J. Crosby
BirdLife International,
Wellbrook Court,
Girton Road,
Cambridge CB3 0NA, UK

Blair Csuti
Idaho Cooperative Fish & Wildlife Research
    Unit,
College of Forestry, Wildlife & Range Sciences,
University of Idaho,
Moscow, Idaho 83843, USA

Frank W. Davis
Department of Geography,
University of California,
Santa Barbara, California 93106, USA

Robert De Wulf
Laboratory of Remote Sensing and Forest
    Management,
University of Ghent,
Belgium

John B. Hall
School of Agricultural & Forest Services,
University College of North Wales,
Deiniol Road,
Bangor, Gwynedd LL57 2WW,
Wales, UK

Julie P. Hawkins
SSC Coral Reef Fish Specialist Group,
c/o Ocean Voice International,
2883 Otterson Drive,
Ottawa, Ontario K1V 7B2, Canada

A.D. Hollander
Department of Geography,
University of California,
Santa Barbara, California 93106–4060, USA

John MacKinnon
Asian Bureau for Conservation,
18E Capital Building,
175–191 Lockhart Road,
Wanchai, Hong Kong

Don E. McAllister
Canadian Centre for Biodiversity,
International Marine Life Alliance,
2883 Otterson Drive,
Ottawa Ontario K1V 7V2,
Canada

Frances Michelmore
PO Box 21472,
Nairobi, Kenya

Ronald I. Miller
Department of Forestry and Wildlife
  Management,
Holdsworth National Resources Center,
University of Massachusetts,
Amherst, Massachusetts 01003, USA

George V.N. Powell
RARE,
Apartado 10165,
San Jose 1000,
Costa Rica

John H. Rappole
National Zoological Park,
Conservation and Research Center,
Front Royal, Virginia 22630, USA

Callum M. Roberts
Eastern Carribean Center,
University of the Virgin Islands,
St Thomas, US Virgin Islands 00802, USA

Steven A. Sader
College of Forest Resources,
260 Nutting Hall,
University of Maine,
Orono, Maine 04469, USA

J. Michael Scott
Idaho Cooperative Fish & Wildlife Research
  Unit,
College of Forestry, Wildlife & Range Sciences,
University of Idaho,
Moscow, Idaho 83843, USA

Frederick W. Schueler
Canadian Centre for Biodiversity,
Canadian Museum of Nature,
P.O. Box 3443, Station D,
Ottawa, Ontario K1P 6P4, Canada

David M. Stoms
Computer Systems Laboratory,
1140 Girvetz Hall,
University of California,
Santa Barbara, California 93106–3060, USA

Preface

The diversity of life is displayed by a diversity of structural and functional elements. Many aspects of this diversity are critical for maintaining the healthy functioning of biological systems both within short and long time scales. Some highly diverse features of nature arise simply from the heterogeneous patterns that comprise the web of nature. Many of these features contribute to the beauty and quality of life. Humans do not yet understand enough about the complexity of nature to distinguish those elements that act to support natural vitality from those elements that contribute exclusively to our experience of beauty and quality in life.

In every region of the world where nature is being preserved, databases and maps are becoming compelling tools for maximizing the likelihood of the future protection of nature. Oftentimes conservation scientists develop unique techniques to map the distribution of the biodiversity elements. These unique approaches are usually tailored to the region of the world where the scientists' work is focused. This book presents accounts of many techniques that are currently being used in different parts of the globe by conservation scientists.

Many different techniques are necessary to handle the differences in data types and data coverages that occur across the globe. Also, a variety of mapping approaches are needed today to strengthen the many diverse critical conservation objectives. These objectives include the identification of the distribution patterns for a species or habitat type and the placement of protected area boundaries. Many of the most advanced techniques that are now being used actively by programs at the forefront of conservation science are presented in these chapters, and their diversity well illustrates the spectrum of database and mapping techniques used today to represent the natural world.

Acknowledgements

I would like to thank several scientists who kindly reviewed manuscripts of chapters that appear herein. I would especially like to thank my editors at Chapman & Hall, Clem Earle and Bob Carling, for their strong editorial and personal support during the production of this book. I would also like to thank Curt Griffin in the Department of Forestry and Wildlife Management at the University of Massachusetts for his support during the final stages of producing the text. Finally, I would like to thank Richard G. Wiegert for his initial insight and support, which made this undertaking possible.

Part One

Introduction

Chapter one

Setting the scene

Ronald I. Miller

1.1 Nature conservation in the world today

A lot of attention is currently focused upon the conservation of nature. For example, an international biodiversity treaty was recently signed by many countries in Brazil (UNEP, 1992), and a compendium that covers many of the known elements of biodiversity was published in association with that treaty (World Conservation Monitoring Centre, 1992). A number of publications have focused upon the most practical and effective strategies for the conservation of nature (e.g. Decker *et al.*, 1991; McNeely *et al.*, 1990; WRI *et al.*, 1992; Soule, 1991); in addition, a recent volume addresses the policy and strategy issues related to the conservation of biodiversity (WRI *et al.*, 1992). The broad scope of biodiversity conservation is well represented in the Global Biodiversity Strategy (WRI *et al.*, 1992, Figure 5). This book is a presentation of some methods currently being used globally to map nature. These methods include prime examples of approaches that employ databases and maps as conservation planning and monitoring tools.

A deluge of information is in circulation in all portions of the management and policy-making arenas. The people in these realms do not have the time to read everything that comes across their desks and, under these conditions, text is not an effective means for communicating important subjects – particularly when an immediate response is required. Maps are powerful vehicles for communicating information. Using only bits of capsulized text, maps permit us, visually, to convey important information about the status, and sometimes the dynamics, of species, habitats and natural resources.

Conservation biology is an interdisciplinary field that incorporates a wide range of disciplines (Temple, 1991). This interdisciplinary perspective is essential for coping with the host of issues that influence the implementation of integrated conservation strategies. A diversity of objectives that are part of these strategies can be met through the use of species and habitat maps.

A focal point of some recent discussions is the contrast between strategies that focus upon species-by-species conservation, ecosystem conservation and biological community conservation (e.g. Hutto *et al.*, 1987; Seal *et al.*, 1992). This book presents approaches used by some of the foremost active conservationists in the world today. The common theme that links these programs is the highly practical character of the approaches that they all employ. The protection of nature in general, rather than

Mapping the Diversity of Nature. Edited by Ronald I. Miller.
Published in 1994 by Chapman & Hall, London. ISBN 0 412 45510 2.

concern for the protection of one natural element versus another, is the principle overriding all these programs. For example, the panda conservation strategy in Chapter 8 is a species-specific project. However, this strategy necessarily must consider a wide range of factors that affect panda populations. These factors include panda biology, distinctive species-specific bamboo flowering cycles, and human population settlement patterns. Each of these elements must be considered for the successful implementation of a practical panda conservation plan. Conservation strategies must be tailored to the circumstances in each specific area of the world. Sometimes this will require conservationists to focus upon species-specific factors and sometimes ecosystem protection factors.

The chapters of this book present many examples of different functions for maps in conservation. These are 'real-world' applications of approaches that use databases and maps to depict the spatial distributions of species, human populations, landscape features, habitats, migration routes, etc. These approaches do not attempt to model structures and processes in the natural world. A recent book presents an excellent overview of current spatial modeling approaches as they can be applied to ecological systems (Hunsaker *et al.*, 1993). Another recent book reviews applications of the Geographic Information System (GIS) to landscape ecology (Haines-Young *et al.*, 1993). The approaches introduced here represent applications developed with the intention of influencing the design, planning and implementation of programs to protect species and habitats.

1.2 Natural features data

On a map boundaries of structural features (i.e. features created by humans) are far more precisely definable than the boundaries of natural features. The boundaries of structural features are discrete and they can usually be precisely located on a map using a combination of point, arc and polygon coordinates. However, the boundaries of most natural features (e.g. species and habitat distribution patterns) are not usually definable in this same way. The movements of animals are not precisely predictable from one moment to the next. The boundaries of habitats change regularly due to the effects of climate, succession, disturbance, etc.

In addition to the potential imprecision of feature boundaries, any single map represents only one of many possible cartographic views of a variable or set of data (Monmonier, 1993). Therefore during the production of species and habitat distribution maps, the authentic depiction of the data should always be carefully considered.

1.2.1 Data variance

A basic characteristic of structural organization in the natural world is the wide variety of climatic, geologic and topographic features. The data that define the status of these ecosystem elements consequently exhibit a particularly wide range of variation. Therefore the variability of ecological data significantly influences the ability of scientists to cogently map natural features.

The data required to produce the maps presented in this volume extend across a broad spectrum of heterogeneity (Figure 1.1). A number of data attributes delineate this heterogeneity including data type, quality, density and precision. The most appropriate mapping technique is often based upon the data intensity that needs to be represented on a map. Consequently, approaches vary across a spectrum between (i) the representation of features on a map with the data registered to coordinate values; and (ii) features on a map presented with only generalized localities and not registered to a coordinate system.

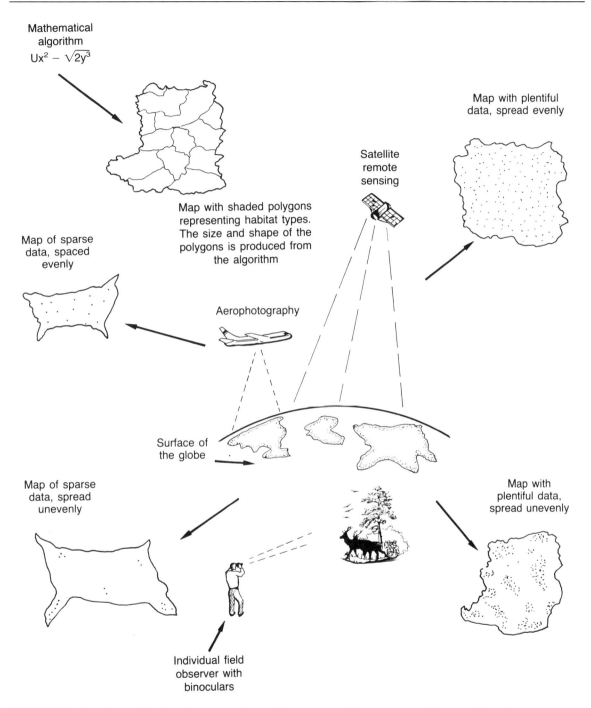

Mathematical algorithm
$Ux^2 - \sqrt{2y^3}$

Map with shaded polygons representing habitat types. The size and shape of the polygons is produced from the algorithm

Map with plentiful data, spread evenly

Satellite remote sensing

Map of sparse data, spaced evenly

Aerophotography

Surface of the globe

Map of sparse data, spread unevenly

Map with plentiful data, spread unevenly

Individual field observer with binoculars

FIGURE 1.1 Species and habitat data are collected from many different sources. Presented here is a spectrum of available map representations for habitat and species distributions. The final map type produced from these data in a given situation is usually dependent upon the available data density and the level of resolution required on the final maps.

5

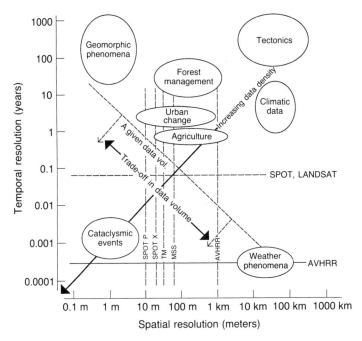

FIGURE 1.2 Spatial and temporal scales for environmental variables and remotely sensed data. Data density decreases from the lower left to the upper right (reprinted with permission from Davis *et al.*, 1991).

The types of data that need to be accessed to produce habitat and species distribution maps vary across a wide spectrum (Figure 1.1). The first distinction to be made is whether the patterns on the map will be produced from data that are predicted from an algorithm or if the map patterns will be produced from observed data that were collected in some capacity (e.g. field observations, aerophotographs, satellite imagery). A consideration of the data intensity and the equivalence of coverage (Figure 1.2) is required in the next stage of map production. The bias and the accuracy of the data also need to be considered in relation to the required final products. The variability of the data will also need to be registered so that the 'roots' of the map data are clearly documented. The procedures and requirements for conservation data documentation are discussed in detail in Chapters 3, 5, 7, 9 and 11.

1.2.2 Data processing

A variety of changes occur during the processing of geographic data for the production of maps. These may include changes to (Davis *et al.*, 1991):

- the datum value (i.e. category, level or magnitude);
- the range of a variable;
- the data precision (higher → lower);
- the spatial or temporal resolution (higher → lower);
- the data type (e.g. numerical → ordinal → categorical);
- the data structure (i.e. tabular → vector → raster); and
- polygon attribute information (i.e. for GIS* data).

The '→' symbols in the above list indicate the

[1] A list of acronyms, and their definitions, is given on page 211.

direction that change will follow if it occurs. In other words, any change that occurs during the processing procedures will necessarily produce lower data precision and lower spatial or temporal resolution. Though the processing changes will probably facilitate analysis, modeling, and/or map production, they will never improve the precision and resolution of the data! A conceptual framework that portrays a lucid spatial context for biodiversity data is presented in Chapter 5.

1.3 The labyrinth of scale

1.3.1 An ecological perspective

Many natural systems exhibit hierarchical organization with nested patterns and processes occurring over a wide range of characteristic space–time scales (Davis *et al.*, 1991). Numerous spatial scale issues therefore need to be addressed when biodiversity maps are produced. For example, a phenomenon may appear homogeneous at one spatial scale but heterogeneous at another (e.g. Goodchild, 1980 as cited in Davis *et al.*, 1991):

(1) the proportions of a region occupied by different land-cover types depend on the spatial resolution of the mapping system;
(2) the distribution and magnitude of slopes on a topographic surface depend on the density of the elevation measurements.

To ensure the creation of valid biodiversity maps, the scale properties of spatial variables should guide the collection, processing and interpretation of the data. These important issues all relate to questions that involve the representation of nature on maps and in databases.

1.3.2 The map scales of natural features

A map organizes a regional picture of an ecosystem or of the status of biodiversity at a single spatial and temporal scale. Bias is frequently encountered during the interpretation, compilation and mapping of the field data used to create this picture on the small map scale. This bias and associated error can be eliminated with the use of standard techniques. For example, the US Environmental Protection Agency and The Nature Conservancy are adopting a hexagon tessellation technique that regionalizes the US into hexagons. It is thought that the storage of data into hexagons is highly effective for analysis and planning purposes (The Nature Conservancy, pers. comm.). A recent study proposes a practical method that quantitatively compares the effectiveness of data storage techniques by comparing land cover classification categories of different maps (Finn, 1993). Up-to-date standardized techniques for identifying and mapping species and habitat distribution patterns in different biogeographic regions are presented in this volume.

A map is a two- or three-dimensional representation of features that may vary in scale across time, space and organization. This is the same variation that influences the patterns generated by ecosystem processes (Levin, 1992). Understanding ecological systems requires a comprehension of how pattern and variability change with the scale of the description (ibid.).

Mapping methodologies developed for presenting species data on the small map scale have evolved at a relatively rapid pace over the past decade (Miller *et al.*, 1989; ICBP, 1992). These techniques present some important mapped elements of biodiversity to planners both in the conservation and in the development community. These maps help to open the eyes of planners to the spatial issues that need to be addressed.

1.3.3 Some scale issues

The representation of natural features on maps is dependent upon the scales of data measurement, data collection and map production. These factors include (Davis *et al.*, 1991):

- the relation of scale to the electromagnetic variation measured by satellite sensors
- the potential measurement scales of maps produced by GIS and remote-sensing technologies
- the influence of data-processing algorithms and flow on measurement scales.

Programs are available that analyze landscape structure in regard to scale. For example, a recently published model samples areas of various sizes that are user-defined and these areas are then used to obtain information about landscape structure at several scales simultaneously (Baker and Cai, 1992).

1.4 Species mapping

1.4.1 An overview

The cartographic representation of species occurrence patterns is a difficult endeavor. Some of the problems are caused by the variability of elements that determine the location of plants and animals at a given point in time. For the purposes of this discussion, a map will represent the distribution of a species within a specified region at one moment in time. Traditionally field biologists have usually not been focused upon recording precise geographic location information. This situation is now changing with the advent of the GIS.

A distinction must be made from the outset between the distribution of species in the real world and the distribution of species as they are portrayed on maps. The distribution of plants and animals on the surface of the globe is regulated by a nearly infinite number of biological and environmental variables. Maps can provide us with very detailed species distribution information, as is produced for scientific monographs (see Chapter 2). Maps can also

provide generalized range information that is less precisely linked to geographic locations (see Chapter 3). However, small-scale maps are not able to give a comprehensive representation of the legitimate complexity of nature.

A variety of species and environmental characteristics influence the spatial distribution of species occurrences in nature. *In today's world, the geographical locations of species occurrences are primarily a composite result of the impact of human civilization on environmental features.* Listed below are examples of biological and behavioral life-history characteristics of species that influence their distribution patterns.

Plants
  dispersal strategies
  reproductive strategies
    asexual
    sexual
  nutrient requirements (including H_2O)
  distribution of critical natural resources.

Animals
  reproductive strategies
    external
    internal
  reproductive behavior
    herds
    leks
    pair-bonding
  territorial behavior
    migratory versus non-migratory species
    defends one territory
    nomadic
  food requirements
  distribution of critical natural resources.

Each of these variables is identified with different aspects of the life cycle of a species. Disturbances due to human encroachment will differentially influence each of these different aspects (Urban *et al.*, 1992). Consequently the effect of human encroachment on distribution patterns will differ considerably from species to species.

1.4.2 Species habitats

Species occurrences are closely linked to the distribution of highly diverse habitats across the surface of the globe. Habitat data types vary considerably across wide response spectra. To illustrate this point, the characteristics of some habitat data derived from published sources for the vertebrate fauna of Venezuela are presented in Table 1.1. These habitat data vary both in the criteria and in the quantitative scales of variation that are used to classify them.

Habitat features hypothesized to influence species distribution patterns can be mapped and analyzed in relation both to individual species and to species richness distribution patterns (Craighead *et al.*, 1985, 1988; Nix, 1986; Freed *et al.*, 1989; Miller *et al.*, 1989). These approaches produce maps derived both from field data and from satellite imagery data. Realism, accuracy and the effectiveness of the depiction of important habitat features need to be carefully considered during the design of these maps. In addition, the definitions, quantifications and variabilities of the elements used to map key habitat features need to be carefully validated (Table 1.1).

New insights into species–habitat relationship patterns are being derived from these new mapping techniques. For example, both a high density of species and a high density of species with small geographical range sizes occur in areas of high habitat diversity (Pagel *et al.*, 1991). These patterns are critical elements for the development of effective conservation strategies.

1.4.3 Map portrayal of species data

The procedures used for the collection and storage of the data are of predominant importance during the mapping process. The *representation* of plant and animal distributions on maps is regulated by the parameters that are used during the map production.

The mapped range of a species is greatly influenced both by the map scale and the data resolution. On the small map scale that covers a broad area, patterns of species distribution are generalized to display the species range across the entire region being shown by the map (e.g. the African elephant distribution – see Chapter 7). On the large map scale that covers a limited area, the depiction of the species range is more closely connected to the actual ground position of the species (e.g. the distribution of a giant panda population – see Chapter 8). Thus the resolution of the data has much less of an influence on the final map depiction of the species range at the small map scale than at the large map scale.

Methodologies for mapping species distribution records evolved slowly until the 1980s, when several studies used small-scale maps to locate and analyze the distribution patterns of rare species in relation to habitats (e.g. Cody, 1986; Feoli and Orloci, 1985; Miller, 1986; Miller *et al.*, 1989). The coarse resolution of data coverage remains a daunting issue in regard to species data. A recent attempt to catalogue and map endemic bird species data on the global scale addresses many of these concerns (ICBP, 1992). This compilation effort effectively deals with many of the issues presented by the diversity of data collected by many different sources using many different collection and recording techniques. Michael Crosby presents the data elements used in this project in Chapter 9 along with some excellent examples of endemic bird distribution maps from several areas in the world.

A number of data resolution issues still need to be addressed to effectively represent species data on the small map scale. It is necessary to establish a range size criterion for mapping species at this scale. Scientific terms related to spatial characteristics need to be defined uniquely at the small map scale. For example,

TABLE 1.1 Examples of available data types needed to document rare vertebrate species distribution in Venezuela. These represent the physical, climatic and vegetational parameters that influence the spatial distribution of rare vertebrate species in Venezuela. The ranges of parameter values were compiled from research accounts of the Rancho Grande region of northern Venezuela (Beebe and Crane, 1947; Test *et al.*, 1966; Steyermark and Huber, 1978)

Parameters	Units of measurement	Range of parameter values
Remotely determined data available from maps, aerial photos, satellite imagery, etc.		
Elevation	meters	0–2765
Climate		
Total precipitation	mm/year	500–2200
Mean monthly precipitation	mm/month	10–600
Number of wet days	days/month	
Temperature	degrees	
Vegetation types		*Elevational ranges*
Cactus/scrub savanna		0–500 m
Deciduous forest		100–900 m
Semi-evergreen forest		400–1100 m
Montane cloud forest		800–2000$^+$ m
Rivers/streams		
Large (16 in Venezuela)	meters across	10 m< <50 m
Small (many)	area drained	
Soils A soil type classification was not accessible at this writing.		
Ground-truthed data acquired during field survey		
Habitats		
Riverine	size, flow rate	
Open canopy		
Closed canopy		
Swamps and ponds	hectares of coverage	
Rocky outcrops/caves	hectares of coverage or number	
Special habitats	hectares of coverage	
Bamboo		
Heliconia		
Leaf litter depth	millimeters	
Disturbed habitat		
Tree falls	number per hectare	
Edges	meters per hectare	
Geologic history		
Geology	age	millions of years
Valleys	width	kilometers of width
	age	millions of years

endemism for birds was defined rigorously by ICBP (1992) at the small map scale. However, maps showing the accurate range of bird species have only become available in South America during the latter part of the 1980s and they are still unavailable for most of continental Asia (ICBP, 1992).

More and more, species and taxon richness maps are providing policy makers with domains upon which to focus. A global perspective for turtle and tortoise densities has recently been created (Kiester *et al.*, 1991). Chapter 10 presents an effective approach for depicting rare coral reef fish distributions around the world.

1.5 Applications of satellite data to natural features mapping

1.5.1 Introduction

Over the past several years, researchers have investigated the validity and accuracy of data collected using satellite technologies. The spatial resolution capabilities of these technologies have significantly improved – AVHRR imagery (1 km^2) – Landsat MSS imagery (80 m^2) – Landsat TM imagery (30 m^2) – SPOT imagery (10–20 m^2) – so that today our view of the earth is improved considerably. In the final chapter I discuss an advanced imagery technology (AVIRIS) that has a potential resolution to provide an accurate and detailed view of the earth's surface.

Satellite data have many potential capabilities for vegetation mapping (e.g. for some examples with MSS and TM see Treitz *et al.*, 1992). However, poor spectral and radiometric resolution produce a low level of detail in these types of imagery. One basic reason for these limitations is that infrared light has a greater wavelength than visible light and this permits less focus in the images. Also variation in the spatial and temporal measurement scales for both field and satellite data restricts the usefulness of these data (Chapter 5). Nevertheless, remotely recorded data can successfully serve as a quantitative ecological baseline for investigating wilderness environments and identifying species habitats (e.g. Craighead *et al.*, 1985, 1988).

1.5.2 Some recent methods

The partitioning of spatial data to produce landscapes is manifest in the edges that emerge from the data itself (Merchant, 1993). Satellite data are most useful for creating representations of landscape cover on the small map scale. Using satellite data, we develop models that describe cloropleth areas as they appear on remotely sensed data (ibid.). The edges shown on small-scale maps are artifacts of the cloropleth mapping process. At present there is no agreed standard method to describe the characteristics and variance of these edges.

To practically analyse land areas on the surface of the globe, scientists have found it necessary to designate land areas as a categorical parameter. Therefore in most mapping studies involving natural features, it becomes necessary either to create an original or to use a pre-existing landcover classification system. During the past 20 years, landcover classification systems have been developed at a variety of scales for many areas on the globe. The process of interpreting the categories between different classification systems is called 'crosswalking'. The crosswalk between landcover classification systems is scientifically a rather subjective process because of both scale and environmental considerations that vary among maps. Nevertheless the crosswalk can be very useful in conservation and development regional planning.

The results of recent research provide a vast array of possibilities for future applications of satellite imagery to vegetation mapping. An extensive series of tests were recently conducted to measure classification accuracy using TM and SPOT imagery data (Wilson and Franklin, 1992). An important conclusion from this study is that different parts of the image may be classified more accurately using different decision rules. The effectiveness of classification approaches was recently measured by comparing a relatively objective approach

(TWINSPAN) with a manual grouping of field plot data into classes (Treitz *et al.*, 1992).

One approach employed extensively with satellite data is to use NDVI to measure vegetation biomass changes on the earth's surface. A recent study concluded that the dynamics of vegetation function are best represented in a vegetation classification system using 'biozones' that are defined by satellite sensor data using the NDVI (Soriano and Paruelo, 1992).

1.5.3 Some new applications of satellite imagery

In recent times the application of satellite data in the natural sciences often focuses upon vegetation change detection (IGBP, 1992a) and measurement of the potential effects of global change (Leemans, R., pers. commun.; Solbrig *et al.*, 1992). Vegetation change detection is usually discernable from a comparison of imagery data collected in different years (e.g. Wilson and Franklin, 1992; Ringrose and Matheson, 1992). However, methodologies for modeling the dynamic nature of the vegetation on the earth's surface are very limited. A recent IGBP report states 'At present there is no mechanism for incorporating the feedback of a changing land surface in a dynamic, interactive way into global models of the physical climate system or of biogeochemical or hydrological cycles. Global vegetation is assumed to be static' (IGBP, 1992b).

The classification of vegetation, even regionally, using satellite imagery data is a somewhat difficult and unwieldy undertaking. One of the reasons for these difficulties is that the spatial and temporal measurement scales for ground samples and remote-sensing data vary considerably (Figure 1.2). The evolution of the satellite remote-sensing technology has far surpassed the application of the imagery data in the natural world. Only today is research beginning

to discover ways to cogently identify diverse vegetation features using satellite imagery data.

1.6 Conservation maps and the GIS

Paper maps are seriously restricted in contrast to maps stored in a GIS (Fletcher and Gibb, 1990). Details are necessarily lost on paper maps because the volume of the original data must be reduced. On paper maps, more accurate presentations and drawings of complex themes are required. Whereby in the GIS the capability of storing multiple map layers that are focused upon individual themes significantly reduces the necessary complexity of the map presentation.

From the outset, geographic information systems were seen as management tools. Computer models for landuse planning provided multiple-uses for forest lands concurrently with timber management facilities (Johnston, 1987). In addition, analyses facilitated by the GIS track relations between rates of natural resource loss and environmental degradation (Johnston *et al.*, 1988). The goals of conservation planning and management are equally well suited to the capability of current computer systems to integrate multiple data types for decision making.

During the process of mapping natural resources, it often becomes necessary to extrapolate or interpolate information about natural resource distribution patterns. The extrapolation is usually done from data collected by scientists at a specific site. The interpolation is usually managed using satellite imagery. A recent study proposes that two requirements exist for the spatial extension of site data: (1) an appropriate statistical model and (2) the availability of a suitable and practical database (Mackey and Bayes, 1990). Conservation policy needs to reconcile the precision of data collected by scientists with the generalized polygons produced from the use of the cloropleth mapping process.

TABLE 1.2 Some of the elements necessary to implement methods for analyzing and identifying priority areas in conservation. Many of these elements will be accessible in either digital and/or paper formats and their accessibility will vary greatly depending on location. An excellent conceptual measurement framework for these elements is presented in Chapter 5

Data types	Data layers	Data input, management and storage	Analysis approaches	Output

Spatial data
Temporal data
Descriptive data
(characteristics of
species and habitats)
	Species data
	Climatic data
	Physiographic data
	Vegetation types
	Soil types
	Geologic data
	Land-use data
	Human impacts data
	Administrative data
		Interpretation of
		various data types
		including:
		– spectral data
		– species field data
		– environmental
		  data
			Spatial modeling
			Statistical modeling
			Gap analysis
			Expert group analysis
				Site selection
				Resource
				  identification

Other data forms		Map digitization
		Database
		  development and
		  packing
		Registration of all
		  spatial data to a
		  single basemap
		  with a unique
		  coordinate system
		  and projection
		Access to required
		  basemap features
		  including:
		– administrative
		  boundaries
		– developed area
		  features
		– natural area
		  boundaries		

1.7 Using databases and maps in biodiversity conservation

Many factors need to be considered for the creation of effective biodiversity conservation strategies and programs. Some of these efforts and their elements are presented in Table 1.2. Species mapping performs several key functions in the conservation process. Maps depicting the distribution patterns of individual species are a key tool for making informed judgements about the conservation status of individual species and for identifying geographic gaps in available conservation data. Critical areas for biodiversity conservation can be identified from analyses of species groups in coordination with habitat patterns (see Chapters 4, 5, 8, 11 and 12).

1.7.1 Some challenges

The identification of priority areas and the analysis of conservation needs is weakened by the non-availability of required information. This is primarily due to a basic lack of

TABLE 1.3 A matrix of some of the factors and costs necessary for the establishment of a basic biodiversity monitoring system. The specifications associated with each individual location will require specific elements not included in this list. Based upon the circumstances, collection of these factors will need to be tailored to each biodiversity monitoring approach as it is implemented

Factors	Costs				
	Purchase	*Setup*	*Upkeep*	*Operation*	*Extended support*
1. *Data requirements*					
Data types					
Data formats					
Necessary comprehensiveness of data coverage					
Required data precision, accuracy, and certainty					
2. *Hardware requirements*					
Computer sizes (i.e. mainframe, mini, PC)					
Computer brands (e.g. IBM, VAX, Apple, etc.)					
Computer attributes					
– memory storage capacity – RAM and ROM					
– graphics capability					
– operating system					
– special feature requirements (e.g. math coprocessor, modem, etc.)					
Plotter type					
Printer type					
3. *Software requirements*					
GIS					
Imagery processing system					
Transfer programs (from the imagery to the GIS)					
Database system					
Spatial analysis software					
Statistical analysis software					
Modeling software					
4. *Mapping requirements*					
Attributes of hardcopy maps (e.g. size, coordinate system, projection, overlay conditions)					
Basemap features					
Precision and accuracy					
Scale requirements					
Degree of generalization permitted					
Map data formats (i.e. hardcopy, digital)					

coherence in the methodologies involved in the collection, storage and analysis of geographically referenced species data. By reviewing elements of species conservation and their association with habitat conservation, I hope that this book will provide a positive influence on current conservation planning strategies.

Species and habitat maps provide the capability for understanding many diverse natural patterns in the world, but the practical application of mapping capabilities abides together with the costs of these programs (Table 1.3). These factors and costs need to be tailored to the unique part of the world in which a specific program is located. In addition, the elements in Table 1.3 are presented only in a summary format and the elements will always need to be explicated for each specific objective within a program.

1.7.2 Some recent mapping approaches

One approach to biodiversity conservation over the past 20 years has been the attempt to identify biodiversity hotspots (Myers, 1988, 1990). For species at a given location, many biogeographic factors (e.g. climate, disturbance) vary unpredictably over time. When some ideal conditions combine, they produce a biodiversity hotspot. The biogeographic factors that contribute to the formation of biodiversity hotspots depend on the number and variability of the biogeographic variables, on differences in the relative importance of these variables to the species, and on the mutual dependences of the species. These are the same biogeographic variables that influence the occurrences of species at a given location over time (Hengeveld, 1990). Once again, the location of biodiversity hotspots in today's world is seriously affected by the impact of human civilization on the environment. A recent consideration of this approach in Africa identifies the complexity of factors needed to be addressed for the identification of biodiversity hotspots (Pomeroy, 1993). Chapters 9, 10 and 11 all include approaches for predicting where biodiversity hotspots are located.

Another conservation strategy being explored today is the protection of taxonomic diversity (Vane-Wright et al., 1991; Faith, 1992). The biogeographic synthesis of systematic and geographic methods provides scientists and politicians with the ability to identify diversity hotspots and to understand the biosphere's evolutionary infrastructure (e.g. biogeographic nodes and tracks – Grehan, 1992). For example, spatial modeling now permits us to analyze historical biogeographic patterns of extinction

using spatial/statistical analysis techniques (Raup and Jablonski, 1993).

1.8 Summary

This book presents a diversity of objectives embodied within contemporary mapping of threatened species and habitats, and many of the factors that motivate the creation of species and habitat maps around the world are incorporated in its chapters. One evident conclusion that emerges from these chapters is that maps are powerful monitoring tools. Maps can provide us not only with important regional information about species and habitat distributions, but also with precise location information about dynamic distribution patterns in relation to landscape features.

References

Baker, W.L. and Cai, Y. (1992). The r.le programs for multiscale analysis of landscape structure using the GRASS geographical information system. *Landscape Ecology*, 7(4), 291–302.

Beebe, W. and Crane, J. (1947). Ecology of Rancho Grande, a subtropical cloud forest in northern Venezuela. *Zoologica*, 32(5), 43–64.

Craighead, J.J., Craighead, L. and Craighead D.J. (1985). Using satellites to evaluate ecosystems as grizzly bear habitat. Grizzly Bear Habitat Symposium, 30 April–2 May, Missoula, Montana.

Craighead, J.J., Craighead, L. and Craighead D.J. (1988). Mapping arctic vegetation in northwest Alaska using Landsat MSS imagery. *National Geographic Research*, 4, 496–527.

Cody, M.L. (1986) Diversity, rarity, and conservation in Mediterranean-climate regions, in *Conservation Biology: The Science of Scarcity and Diversity* (ed. M.E. Soule), Sinauer Associates, pp. 122–52.

Davis, F.W., Quattrochi D.A., Ridd, M.K. *et al.* (1991). Environmental analysis using integrated GIS and remotely sensed data: Some research needs and priorities. *Photogrammetric Engineering and Remote Sensing*, 57, 689–97.

Decker, D.J., Krasny, M.E., Goff, G.R. *et al.* (eds) (1991) *Challenges in the Conservation of Biological Resources: A Practitioner's Guide* (1st edn), Westview Press Inc., Boulder, Colorado, and Oxford, England.

Faith, D.P. (1992). Conservation evaluation and phylogenetic diversity. *Biological Conservation*, 61, 1–10.

Feoli, E. and Orloci, L. (1985) Species dispersion profiles of anthropogenic grasslands in the Italian Eastern Pre-Alps. *Vegetatio*, 60, 113–18.

Finn, J. (1993) Use of the average mutual information index in evaluating classification error and consistency. *Journal of Geographical Information Systems*, 7 (4), 349–66.

Fletcher, J.R. and Gibb, R.G. (1990) *Land Resource Survey Handbook for Soil Conservation Planning in Indonesia*. DSIR Land Resources Scientific Report 11, 127pp.

Freed, L.A., Cann, R.L. Scott, J.M. *et al.* (1989) Integrated conservation strategy for Hawaiian forest birds. *Bioscience*, 39(7), 475–9.

Grehan, J.R. (1992) Biogeography and conservation in the real world. *Global Ecology and Biogeography Letters*, 21, 96–7.

Haines-Young, R., Green, D.R. and Cousins, S.H. (eds) (1993) *Landscape Ecology and GIS*, Taylor & Francis, London, New York and Philadelphia, 288pp.

Hengeveld, R. (1990) *Dynamic Biogeography*, Cambridge University Press, Cambridge, UK.

Hunsaker, C.T., Nisbet, R.A. Lam, D. *et al.* (1993) Spatial models of ecological systems and processes: The role of GIS, in M. Goodchild, B. Parks and L. Steyaert (eds), *Geographic Information Systems and Environmental Modeling*, Oxford University Press, New York.

Hutto, R.L., Reel, S. and Landres, P.B. (1987) A critical evaluation of the species approach to biological conservation. *Endangered Species Update*, 4 (12; October), 1–4.

ICBP (1992) *Putting Biodiversity on the Map: Priority Areas for Global Conservation*, International Council for Bird Preservation, Cambridge, UK.

IGBP (1992a) *Improved Global Data for Land Applications: A Proposal for a New High Resolution Data Set. A Study of Global Change*, Report of the Land Cover Working Group of IGBP-DIS, (ed. John R.G. Townshend), IGBP Report Number 20, IGBP, Stockholm.

IGBP (1992b) *Global Change and Terrestrial Ecosystems: The Operational Plan. A Study of Global Change*, (ed. W.L. Steffen, B.H. Walker, J.S.I. Ingram and G.W. Koch, with contributions by J. Goudriaan, T. Hirose, J.H. Lawton *et al.*), IGBP Report Number 21, IGBP, Stockholm.

Johnston, J.P., Detenbeck, C.A., Bonde, N.E. and Niemi, G.J. (1988) Geographic Information Systems for cumulative impact assessment. *Photogrammetric Engineering and Remote Sensing*, 54 (11), 1609–15.

Johnston, K.M. (1987) Natural resource modeling in the Geographic Information System environment.

Photogrammetric Engineering and Remote Sensing, 53 (10), 1411–15.

Kiester, A.R., Iverson, J.B. and White, D. (1991) Turtle and tortoise conservation: A global perspective from biogeography. Annual Meeting of the American Society of Herpetologists and Ichthyologists, July 1991.

Leemans, R. (1991) The effects of climate change for natural ecosystems in China. Unpublished manuscript, Global Change Department, National Institute of Public Health and Environmental Protection, Bilthoven, The Netherlands.

Levin, S.A. (1992) The problem of pattern and scale in ecology. *Ecology*, 73 (6), 1943–67.

Mackey, B.G. and Bayes, T. (1990) A modelling framework for the spatial extension of ecological relations in vegetation studies. *Mathematics and Computers in Simulation*, 32, 225–9.

McNeely, J.A., Miller, K.R. Reid, W.V. *et al.* (1990) *Conserving the World's Biological Diversity*, IUCN, Gland, Switzerland.

Merchant, J.W. (1993) Characterizing landscape structure using satellite image data. *Environmental Information Management and Analysis: Ecosystem to Global Scales*. An international symposium sponsored by the US National Science Foundation, 20–22 May 1993, University of New Mexico, Albuquerque, New Mexico.

Miller, R.I. (1986) Predicting rare plant distribution patterns in the Southern Appalachians of the southeastern USA. *Journal of Biogeography*, 13, 293–311.

Miller, R.I., Stuart, S.N. and Howell. K.N. (1989) A methodology for analyzing rare species distribution patterns utilizing GIS technology: The rare birds of Tanzania. *The Journal of Landscape Ecology*, 2(3) 173–89.

Monmonier, M. (1993) *Mapping It Out*, The University of Chicago Press, Chicago and London. 301pp.

Myers, N. (1988) Threatened biotas: Hotspots in tropical forests. *Environmentalist*, 8, 187–208.

Myers, N. (1990) The biodiversity challenge: Expanded hotspots analysis. *Environmentalist*, 10, 243–56.

Nix, H.A. (1986) A biogeographic analysis of Australian elapid snakes, in *Atlas of Elapid Snakes of Australia* (ed. R. Longmore), Australian Flora and Fauna Series, no. 7. AGPS, Canberra, pp. 4–15.

Pagel, M.D., May, R.M. and Collie, A.R. (1991) Ecological aspects of the geographical distribution and diversity of mammalian species. *American Naturalist*, 137(6), 791–815.

Pomeroy, D. (1993) Centers of high biodiversity in Africa. *Conservation Biology*, 1(4), 901–7.

Raup, D.M. and Jablonski, D. (1993) Geography of end-cretaceous marine bivalve extinctions. *Science*, 260, 971–3.

Ringrose, S. and Matheson, W. (1992) The use of

Landsat MSS imagery to determine the aerial extent of woody vegetation cover change in the west-central Sahel. *Global Ecology and Biogeography Letters*, **2**, 16–25.

Seal, U.S., Flesness, N.R. and Foose, T. (1992) Endangered species: Anachronism or essential element in protected areas management. *IV Congreso Mundial de Parques Nacionales y Areas Protegidas, Caracas, Venezuela*, 10–21 February 1992, Book Two.

Solbrig, O.T., van Emden, H.M. and van Oordt, P.G.W.J. (eds) (1992) *Biodiversity and Global Change*, IUBS Monograph No. 8. International Union of Biological Sciences, Paris.

Soriano, A. and Paruelo, J.M. (1992) Biozones: vegetation units defined by functional characters identifiable with the aid of satellite sensor images. *Global Ecology and Biogeography Letters*, **2**, 82–9.

Soule, M.E. (1991) Conservation tactics for a constant crisis. *Science*, **253**, 745.

Steyermark, J.A. and Huber, O. (1978) *Flora del Avila: Flora y vegetacion de las montanas del Avila, de la Silla y del Naiguata*. Publicacion Especial de la Sociedad Venezolana de Ciencias Naturales bajo los auspicios de Vollmer Foundation y Ministerio del Ambiente y de los Recursos Naturales Renovables, Caracas, 971pp.

Temple, S.A. (1991) Conservation biology: New goals and new partners for managers of biological resources, in *Challenges in the Conservation of Biological Resources: A Practitioner's Guide*, (1st edn; ed. D.J. Decker *et al.*), Westview Press Inc., Boulder, Colorado, and Oxford, England, pp. 45–54.

Test, F.H., Sexton, O.J. and Heatwole, H. (1966) *Reptiles of Rancho Grande and Vicinity, Estado Aragua, Venezuela*. Miscellaneous Publications No. 28, Museum of Zoology, University of Michigan, Ann Arbor.

Treitz, P., Howarth, M.P., Suffling, J.R.C. and Smith, P. (1992) Application of detailed ground information to vegetation mapping with high spatial resolution digital imagery. *Remote Sensing of the Environment*, **42**, 65–82.

UNEP (1992) *Convention on Biological Diversity*, 5 June 1992, NA-92–7807, UNEP, 24pp.

Urban, D.L., Hansen, A.J., Wallin, D.O. and Halpin, P.N. (1992). Life-history attributes and biodiversity. Scaling implications for global change, in *Biodiversity and Global Change* (ed. O.T. Solbrig, H.M. van Emden and P.G.W.J. van Oordt), IUBS Monograph 8, IUBS, Paris, p. 224.

Vane-Wright, R.I., Humphries, C.J. and Williams, P.H. (1991). What to protect? – Systematics and the agony of choice. *Biological Conservation*, **55**, 235–54.

Wilson, B.A. and Franklin, S.E. (1992). Characterization of alpine vegetation cover using satellite remote sensing in the Front Ranges, St. Elias Mountains, Yukon Territory. *Global Ecology and Biogeography Letters*, **2**, 90–5.

World Conservation Monitoring Centre (1992). *Global Biodiversity: Status of the Earth's Living Resources*, Chapman & Hall, London.

WRI, IUCN and UNEP (1992). *Global Biodiversity Strategy: A Policy-makers' Guide*. A report produced in consultation with FAO and UNESCO, 35pp.

Part Two

A Medley of Contexts for Mapping Species Data

Perhaps for thousands of years, naturalists throughout the world have been collecting data related to the location of species. Many disparate methods have been used to describe these locations. The recent development of the capability to perform computerized mapping with the Geographic Information System (GIS) has effected a change in the methods used to store and represent species distribution information. A wide range of these recent approaches for producing maps and databases from species observation data are presented in this part.

Maps showing the diversity and abundance of plant and animal species provide a valuable tool for environmental planning and management. Such maps help pinpoint where species are located so that both the species and their surrounding environs can be better protected and managed. These maps can serve generalists and experts alike. The maps link geography and ecology and are readily accessible to economists, planners and experts in other disciplines, as well as government officials and the general public.

Monographs are scientific publications that thoroughly document species characteristics. In Chapter 2 a detailed description of the data collection and representation techniques used to produce maps of plant distribution data for monographs is presented. As an illustration, data are presented for several tree species that were not exhaustively studied in the past but are recognized today as important natural resources. The procedures used in monographs to compile species distribution information from many sources and to produce a reliable distribution map are described. Techniques are also presented for accessing sources of environmental data (e.g. terrain, climate and soils) to represent the associated patterns of variation in species distributions. The mapped distributions of five tree species are presented in Chapter 2 to illustrate the range of data variation. The variability of the species distribution patterns and the environmental factors are then related to map production. The chapter urges a central role for distribution maps in the context of monograph production. The implications and uses of these maps in the management and conservation context are then also considered.

Chapter 3 presents a study conducted at the

World Bank in 1988–9 involving the mapping of threatened animal species distribution patterns in Madagascar, a country with abundant unique and rare species. Methods are introduced for geographically interpreting imprecise species field notes and for producing draft maps that display threatened species patterns across the landscape. The boundaries in these maps represent the approximate ranges where the species are most likely to be found. The species location data are then successively refined through consultations with experts with first-hand knowledge of the area under study. The final maps are then rectified as a consequence of the guidance provided by the field experts. This same consultative technique is now being used successfully by Conservation International in several parts of the world (Tangley, 1992). This is proving to be a valuable and reliable method for coping with scientific uncertainty. The chapter then discusses ways to make species maps as reliable as possible.

Chapter 4 presents some up-to-date methods used by the Gap Analysis program in the United States (Kareiva, 1993; Pennisi, 1993) to model vertebrate distribution patterns. Spatial models are used to map the breeding distributions of 366 species of amphibians, reptiles, birds and mammals in Idaho, USA. The models are based on the known range limits of each species and their association with mappable environmental features. The models are applicable to regional planning and analysis. Similar Gap Analysis state-wide models are being developed across the US and they will provide detailed distribution models of vertebrates useful for regional or national conservation planning and analysis.

Up-to-date species range maps have wide applications. They can guide Environmental Action Plans* and conservation strategies in developing countries. They can help to improve the design of national parks and can ensure that existing parks and protected areas successfully protect species. These maps can help identify and evaluate the environmental impact of development projects, and improve project design and management. Most broadly, they can alert people to the location of environmentally sensitive areas with important biodiversity elements.

References

Kareiva, P. (1993) No shortcuts in new maps. *Nature*, 365, 292–3.

Pennisi, E. (1993) Filling in the gaps. *Science News*, 144, 248–51.

Tangley, L. (1992) *Lessons from the Field, I: Mapping biodiversity*, Conservation International.

---

* These are plans devised by developing countries and funded by the World Bank to support environmental protection.

Chapter two

Mapping for monographs: Baselines for resource development

John B. Hall

2.1 Introduction

The last two decades have witnessed an explosive increase in interest in a wide range of tree species never previously acknowledged as resources on a global or regional scale. This trend has arisen from the emergence of tree-growing strategies outside the formal forestry and horticulture sectors as awareness of the significance of trees as key rural resources has spread. Land use initiatives involving collaboration between rural communities and expertise representing government departments or development agencies are a consequence. Increasingly, these are progressing toward independent but informed actions conceived and executed entirely by rural people. This process depends upon the collection and distribution of information about the tree species. Monographs are comprehensive documentary compilations produced to meet that need.

Much of the impetus behind today's relatively amicable liaison between local people and external or governmental institutions originated with the 1978 World Forestry Congress in Djakarta, Indonesia, and scattered initiatives preceding it – notably under US National Academy of Sciences and FAO auspices. Inspirational publications (e.g. National Research Council, 1975, 1977) were complemented with numerous others in the years following the Djakarta congress, at both the international (e.g. National Research Council, 1980, 1983a; Baumer, 1983) and national (e.g. Panday, 1982; Forest Division, 1984; Teel, 1984) levels.

Seeking to encourage action over much of the tropics and subtropics, stress in the early documents was on publicizing a wide range of potentially useful new tree resources, each being necessarily somewhat superficially treated – in a few hundred words of generalized comment. These short commentaries ('profiles')

Mapping the Diversity of Nature. Edited by Ronald I. Miller.
Published in 1994 by Chapman & Hall, London. ISBN 0 412 45510 2.

were successful in their primary aim, promoting interest in tree species previously disregarded in forestry and horticulture. At the same time, however, those preparing them were left in no doubt that relevant, readily accessible information was limited and far from adequate as a basis for devising appropriate management practices. Until the 1980s, instances of in-depth knowledge of what were effectively members of a new generation of economic tropical tree species were rare. The principal exception was *Leucaena leucocephala* (Lam.) De Wit, for which a comprehensive account became available in 1977 (National Research Council, 1977). This was facilitated by the important prior work of Dijkman (1950) and Oakes (1968). A subsequent series of documents (e.g. National Research Council, 1983b, 1983c) treated additional species in comparable detail. But even today, fewer than 20 species out of hundreds of promising candidates (von Carlowitz, 1986) are covered in depth. Even these documentations vary considerably in approach, scale and authority. Some (e.g. Mann and Saxena, 1980; Withington *et al.*, 1987) lack the unity of a single team of authors and the cohesion and comprehensiveness imposed by executive editing. Others (e.g. Bonkoungou, 1986, 1987) concentrate on a restricted part of the species' range and lack geographical balance. Mapping is neglected (e.g. Habit, 1981) or severely limited in resolution in several of these coverages. This chapter urges a central role for distribution maps in the monograph context and examines the procedure whereby such maps are compiled and used.

Individually mapping known localities for a species throughout its range generates a unified, standardized and comprehensive record of its geographical status. The result is often considered to be an end in itself, as with the series of maps issued as *Distributiones Plantarum Africanarum* by Jardin Botanique National de Belgique. However these maps have much to

contribute as a resource that elucidates species ecological relations and morphological trends. How effectively this role can be fulfilled by these data depends on the reliability and coverage of mappable data sources. This also depends on the way in which subsequent interpretation is implemented.

2.2 Assembling the primary data set

The most widely used sources of information can be grouped into five broad categories. The relative importance of each varies with the target species and the stability of its taxonomic circumscription.

2.2.1 Herbarium material

It is frequently considered that an acceptable distribution map is based strictly on herbarium material. In practical terms this is unrealistic, and assumptions that such maps will be error-free are unjustified. Nevertheless, information from the herbarium is of special value because permanently preserved voucher material can be physically examined, re-examined on subsequent occasions, and any reservations about identification noted. Further, the notes provided by the collector provide insight into site conditions, associated species, prominence, stature and uses. However, to locate and examine all herbarium material of a widely distributed taxon is too time-consuming a process to be feasible. Numerous herbaria are registered with the International Bureau for Plant Taxonomy and Nomenclature. Hepper and Neate (1971), for example, refer to over one hundred herbaria simply with respect to holdings of West African material. In any case, there is no consistency in the operations of different herbaria – particularly with regard to resources, engagement of specialist taxonomic expertise and the designated service role. Staffing constraints and unavailability of current and relevant taxonomic literature often produces ignorance of name changes and incorporation of mis-identified specimens. Even in major international herbaria, specimens in critical groups may remain wrongly assigned pending active flora work involving the country of origin. Information from a herbarium, as with that from other sources, cannot be used indiscriminately. Account must be taken of likely nomenclatural misapplications and measures should be taken to detect and exclude these.

2.2.2 Taxonomic literature

Literature citing voucher specimens and geographical localities is a major source of data for mapping. However, consistency between literature and maps in the taxonomic circumscription is essential. The main type of publication relevant is the flora. Regional or national floras (i.e. most volumes are already issued) exist for much of the world and cover the majority of tree species. Unfortunately not all include citations of specimens and some cite examples without specifying their origin. Area floras generally indicate whether any synonyms are in frequent local use. Usually an exhaustive synonymy for species with ranges extending beyond the limits of the flora area is not provided. Appreciation of the synonymy enables the incorporation of correct species names in maps. Maps can thus be produced with information from literature where the same species was considered under a different name. Area floras are of further value as the most convenient means of ascertaining the range of a species in general terms. Frequently, for economy of space, citations are restricted to a few 'representative' specimens that are selected on the basis of their 'quality' rather than to achieve comprehensive geographical coverage. This usually necessitates references to information from other sources as the map is developed.

2.2.3 Inventories

In herbaria, there is often a reluctance to preserve material that is not accompanied by fruits or flowers. As a consequence, many species (e.g. some widespread and frequent) that have relatively inaccessible reproductive shoots or that flower only at widely spaced and irregular intervals are under-represented. Other species feature rarely in collections because they are extremely familiar and field identification is unsupported by voucher specimens. Some species, among which palms are prominent, are rarely collected because of inconvenient morphological characteristics (e.g. large, robust leaves; long, rigid spines; irritant or stinging hairs). Reliance on cited specimens alone therefore does not facilitate a satisfactory appraisal of the distribution. In such cases, incorporating data gathered during forest inventory activities may improve the picture considerably. Inventory records also provide insight into a species' abundance (e.g. individuals per hectare (ha)), usually for a number of size classes, as well as overall.

Set against these advantages are a number of limitations associated with inventory data. Summarized information is not widely circulated and can normally be consulted only through direct access to forest service archives. Also, while it is often acknowledged that good inventory practice provides for the collection of voucher herbarium material, this is seldom undertaken. As a result, it is extremely difficult to check identifications and possible confusions attached to closely allied species, which may lack recognized economic potential. Older sets of inventory data are particularly subject to this weakness. In these older data sets, this problem is compounded by the later redefinition of species limits. Nevertheless, the assignment of skilled and trained tree identifiers to inventory teams, especially in recent years, has minimized the extent of identification difficulties. The main value of inventory data for species mapping is in the closed high forest regions of the tropics. Since savannah vegetation types have little appeal as timber sources, most savannah tree species are disregarded in formal inventories.

2.2.4 Ecological literature

Ecological literature is a useful source of distributional information supplementary to that retrieved from herbaria or taxonomic literature. As with the inventory data, the overall significance of this source depends on the species in question. Ecological literature represents a more extensive mass of data than the inventories and it covers all major vegetation formations comprehensively. Most ecological literature focuses upon the plant community rather than the individual species. This is advantageous in throwing light on the role of the species within the community, but species names are generally excluded as keyword descriptors. Consequently, for the purposes of species mapping, relevant ecological papers are difficult to trace through computerized database search facilities. To make use of this source, familiarity with ecological accounts of the area for the species under consideration is necessary unless convenient bibliographic documents (e.g. Reilly, 1976, for Sabah; White, 1983, for Africa) can be traced.

Within the ecological literature, two forms of publications of particular interest are the vegetation map (primarily at national or a more local level) and the phytosociological survey. Mapping terms seldom relate directly to named species in moist forest vegetation – inventory data are of more use. In contrast, in savannah and more arid regions mapping units are often defined by prominent tree species. This prominence reflects gregariousness or low diversity within a thinly dispersed tree stratum. Maps are never comprehensively supported by voucher specimens and mapped information on species in critical groups should be rejected if there is doubt about data reliability.

Published phytosociological surveys usually incorporate a list of constituent species for each vegetation type encountered and it is customary for a taxonomic specialist to have been involved in the work. This involvement is normally reflected in a set of voucher specimens cited in the species lists, enabling identifications to be checked, if desired. Such surveys tend to be sample-based and on a finer scale than inventory exercises. These surveys incorporate quantitative appraisals of abundance. In addition, phytosociological surveys provide for recording of environmental information which may be helpful in distribution map interpretation.

2.2.5 Personal observations

It is possible to amass first-hand distributional information (on either a systematic basis or as incidental records of occurrence) when opportunities arise for field activity within the range of a species. Funding specifically to support field surveys of distribution is unlikely *in any part of the world!* However, a distributional aspect can complement field touring organized primarily for other purposes (e.g. seed collection, market surveys). Often these data are intended to augment information on distribution derived from other sources. Therefore, when these data are collected, it is crucial that the recorder is fully competent to distinguish the target species from any close relatives (i.e. capable of identification at every season and at every stage in the plant's life cycle).

2.3 Complementary data sets

Complementary data sets comprise information, amenable to representation in map form, that allows a species distribution pattern to be predicted in terms of external influences. Three categories of information traditionally have been useful for this purpose and are most readily available as complementary data. These data categories are terrain, climate and soil.

2.3.1 Terrain

Topographic maps are readily available on continental and multinational scales in atlases. At more convenient scales, they are available as published international map series or topographic maps of individual countries. At scales in the range 1:1 000 000 to 1:5 000 000, contours are likely to be displayed at 100, 200 and 500 m intervals. Above this elevation, contours are presented at 500 m vertical intervals.

2.3.2 Climate

Some continental scale maps of climatic variables are available, but compilations of agroclimatological (FAO, 1984, 1985, 1987) or more conventional meteorological data (Meteorological Office, 1959, 1966, 1983) are more convenient. Access to tabulated information for a wide range of parameters (e.g. precipitation, temperature, humidity, wind speed/wind run and radiation) provides options for the derivation of more complex climatic parameters from such sources, as well as quantification of conditions at specific locations. National meteorological documentation is also often available and these data may clarify local climatic patterns in heterogeneous terrain.

2.3.3 Soil

Unlike climatic maps, soil maps* do not invite interpolation and the resolution must be taken into account when they are related to species distribution maps. Boundaries between soil mapping units are often sharp. Omission of an occurrence of a particular soil type in an area may suggest spurious association with what is really a contiguous soil type. Species associated with habitat conditions that are often linear in configuration, notably riverain species, are particularly subject to this distortion. Patterns may be determined for species that occur extensively, if they are associated with particular mapping units characterized by a widely distributed, dominant soil type.

The ecological literature can usually be used to ascertain if the coarseness of mapping resolution distorts a relationship. Close relationships between a species and specialized soils of restricted extent (i.e. the soil types that maps under-represent) is usually noted during vegetation survey. In addition, this is underlined by consistent independent remarks from localities throughout the range. Awareness of this limitation is evident in the texts for the FAO–UNESCO maps which tabulate 'associated soils' for each unit depicted on the maps. Therefore, it is advisable to consider very carefully any studies that describe pronounced small-scale heterogeneity reflected in an intricate mosaic or pattern of sharply contrasting soil types.

2.4 Procedural considerations

2.4.1 Map production

Once all available specimens and records of the occurrence of a species are located† the geographical coordinates for as many as possible are needed. This requirement is seldom unclear with wide-ranging species.‡ Material without records of the source localities (e.g. particularly in older collections) or with remarks relating to large administrative units rather than specific locations is often incorporated into herbaria. Some early collections received their original labels at major administrative centers from where intermediaries forwarded them to herbaria. Names such as 'Lagos' and 'Zanzibar'

---

* A comprehensive source of soil data is the FAO–UNESCO Soil Map of the World. This presents areas where 26 groups of major soils occur world wide at a scale of 1 : 5 000 000.
† A comprehensive and representative sample of these occurrence records will usually suffice.
‡ Often a proportion of the records and specimens for these species do not contribute information to the final map.

were used to cover enormous areas accessed through these places. Such practices are familiar to the staff of herbaria that hold such material and these naming difficulties are taken into account. Today, anomalies of this type are apparent when a draft map is completed and they are removed if doubts about reliability remain.

More usually, the problem is to translate a place-name into geographical coordinates. For some parts of the tropics there are convenient tools for this in the form of catalogues of collecting localities (e.g. Bamps, 1982; Polhill, 1988) or gazetteers that are incorporated in ecological texts or floras (Letouzey, 1968). General national gazetteers can also be used for identification of coordinates. These are published as international initiatives (e.g. the gazetteers of the US Board on Geographic Names) and in many countries by national survey departments to complement standard topographic map series (e.g. Dadson, 1965). Place-name ambiguities often occur (e.g. places at different locations bearing an identical name). Identification of the correct place-name requires research into the context of a report or collection. Clarification of a place-name can be derived from a collector's general collecting patterns and other activities.

Early collections are often more difficult to associate satisfactorily with geographical coordinates. In these cases place-names may have changed, the collector may have provided a written form of an oral communication, and the settlement that yielded the name may have disappeared. For the more prolific early collectors, however, attempts have been made to summarize activities by accounts of itineraries supported by maps (e.g. Wickens, 1972 for Georg Schweinfurth; Bamps, 1975 for G.W.J. Mildbraed). Much literature relevant to solving problems of this nature is conveniently listed in recent historical accounts of botanical exploration (e.g. Friis, 1992).

Attempts to clarify locations directly from maps are tedious but usually necessary when gazetteers or lists of collecting localities are not available but the source country is known. Maps at scales >1:250 000 are rarely convenient because so many sheets are needed and some countries still lack complete coverages in such detail. For many woody species, a significant proportion (and frequently the great majority) of collections and reports relate to national forest estates. These areas invariably require close inspection of maps that name and locate forest reserves. When they are accessible, there is obvious advantage in consulting maps contemporaneous with the collecting era. It should be noted that early maps are not all based on the Greenwich Meridian used today and exceptions need reinterpretation of coordinates.

Final maps at the continental scale usually depict ca. 0.5–2 cm to represent 500 km. On these maps, resolution is inevitably restricted and location to within 15–30′ of latitude or longitude usually suffices. In the *Distributiones Plantarum Africanarum* maps, for example, resolution is to the nearest 1°, symbols that indicate presence are centered within 1° × 1° map squares. Sometimes the use of GIS techniques that permit superimposition of locality information produce inappropriate results; for example, the resolution underlying continental scale maps of soils and climatic factors is very restricted. Interpolated lines for climatic patterns and soil mapping units represent generalized areas only credible within a range of ca. 400 km^2.

2.4.2 Map interpretation

A preliminary inspection of the map ascertains compactness and continuity of occurrence in relation to the land surface concerned. For an essentially continuous distribution, a major need is to establish if there is a discernible departure from the conditions within the range in the regions beyond the periphery. The change associated with different sections of the periphery may not always involve the same factors. For example, occurrences of frost and inadequate rainfall may limit the distribution in different periphery locations. After a break in the distribution, suitable conditions may recur. If so, the presence of a barrier to spread is indicatcd and its nature should be examined. When a range is clearly in discrete portions (i.e. disjunct), changes concerning the intervals between the disjunct portions should be investigated similarly and the process should be extended to lacunae within the range. A fragmented distribution with an archipelago-like spatial configuration usually indicates a sporadic pattern of variance in one or more environmental factors. If linear configurations are revealed then linear environmental features rather than progressive changes should be considered as controlling factors. Subsequent action investigation should vary according to whether relations are sought with parameters mapped on an isoline, a mosaic/point or a linear basis.

(a) Isoline parameters

In areas where mountain ranges rise out of a generally low landscape, it is usual to check first if occurrences of the species center on land at high elevation. For montane species with a restricted latitudinal range, a well-defined relationship with an altitudinal threshold may be readily detected (Figure 2.1). As ocean temperatures become less consistent at higher latitudes, elevation is no longer a convenient substitute for temperature. Temperature is not as well documented a parameter as elevation, but it can be used effectively when it is identified with point localities.

After determining any relationship between elevation and species distribution, rainfall is next considered. For widely distributed species, rainfall isolines seldom yield the distinct relationships produced by elevation contours. For example, in sites receiving low rainfall, the water supply may be augmented from other

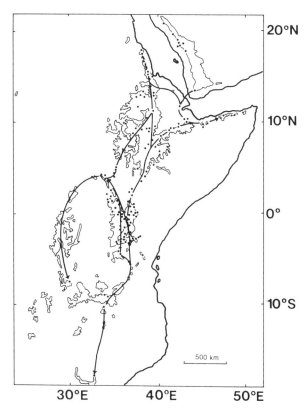

FIGURE 2.1 An example of a distribution related to an isoline parameter. Occurrences (dots) of *Juniperus procera* Endl. and land surfaces ≥1500 m elevation (delimited by the 1500 m contour). *J. procera* is considered to have a fragmented distribution. Heavier lines indicate postulated migration routes from Eurasia into eastern Africa.

sources including: occult precipitation, groundwater, major rivers, and particularly in coastal areas, high atmospheric humidity. Similarly, species widely distributed where rainfall is below a certain threshold penetrate regions of higher rainfall in some circumstances. The sites that are colonized are those where soil conditions are less favorable than across the area generally. Examples are sites with extended periods of waterlogging or where the soil is shallow with prolonged periods of water stress – in areas where the latter arise from land degradation, species from drier zones are generally aggressive colonizers. Usually, in these instances, interactions with factors other than climate or major rivers are involved. Subsequent explanations of this pattern require a review of information provided with the record used to locate the occurrence. Mankind's activities, particularly in the livestock context, frequently generate such effects, with species of drier zones colonizing degraded landscapes in wetter ones, especially along cattle trails.

Once determined, relevant isolines are incorporated in the distribution map to show coincidence with the typical limits of distribution. Anomalous occurrences are then identifiable as those outside the main part of the range.

(b) Mosaic/point parameters

Mosaic/point parameters cannot be analyzed using isoline approaches. In the case of soil types (a mosaic parameter), this is because of the intricacy of the heterogeneity. There are numerous stations that record rainfall, and rainfall generally changes progressively over long distances. Rainfall maps are created by inferring isohyet alignments from a distribution of rainfall records. Other climatic parameters are recorded at many fewer places and often change appreciably over short distances.

The relation between a species distribution and soil types must be analyzed cautiously in view of the problems of resolution already mentioned. A species is commonly tolerant of a range of soil types, notwithstanding stronger relationships with certain varieties. To show such relationships, the symbol denoting occurrence is varied to correspond with the soil type variation (Figure 2.2). On a more local scale where occurrence is quantitatively evaluated, symbols showing the species' prominence are superimposed onto a soils map (e.g. Hall and Bada, 1978).

Point factors can easily be represented in map form and compared with the species distribution map. For climatic variables, points represent the locations of synoptic meteorological stations. At the outset, all such stations within the ranges of the species (i.e. latitude and longitude) are plotted. The information from these stations indicates conditions prevailing where the species grows. However, a relationship of this kind does not necessarily indicate that the extent of the range depends only on these conditions. To determine if this is the case, it is necessary to consider also the climatic information in adjacent areas outside the range. Contrasts are then sought between the area within the range and the area outside of it. Further action depends on whether the distribution of the species is essentially continuous or archipelago-like.

Given a continuity of range, the boundary is a threshold defined by any limiting factor that is consistently lower or higher than the values within the range. If comparable numbers of climatic stations inside and beyond the species range are accessed, then the strength of the relationship between a species distribution and a climatic parameter can be explored in more depth using formal statistical tests of association. Most simply, this requires the concordance of each meteorological station with a state for the variable (e.g. presence or absence of frost) and with a value that indicates whether a location is within or outside the range (Figure 2.3; Table 2.1). Where there are enough data points, a simultaneous association with levels of additional parameters can enter the process. The clarity of the location can be enhanced, for example, when a explanation for a range limit is reached with confidence. Then extraneous information can be deleted from the distribution map, leaving only points of occurrence and the interpolated line.

The interpretation of an archipelago-like distribution is usually complicated by the effects of a highly heterogeneous environment. Temperature tends to be a key element for montane species but it is modified by lapse rates and

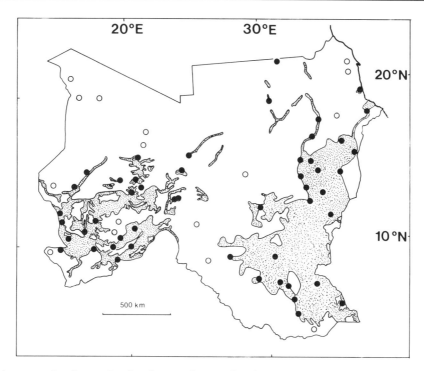

FIGURE 2.2 An example of a species distribution that is related to a mosaic parameter (i.e. soil). Depicted are the occurrences in Sudan and Tchad of *Acacia seyal* Delile var. *seyal*. Also shown is the dominant presence of vertisolic and associated basin soils (stippled) (from FAO–UNESCO, 1977). Expanses of these soils are sufficiently extensive to feature in continental (i.e. small) scale maps only in the range sections shown. Infilled circles denote the presence of this species on vertisolic and associated soils. Open circles denote the presence of this species on other soils.

mass heating (*massenerhebung*) phenomena in mountain terrain. Accordingly, only data from meteorological stations in close proximity to recorded occurrences of the species can be used as guidance to the prevailing conditions. A maximum interval of 25 km between a species location and a meteorological station is suggested under these conditions (also, only locations with known elevations should be used). Each of the relevant stations will yield the baseline temperature data for the species occurrence site. Application of an appropriate temperature lapse rate to the difference in elevation between the two sites will produce a temperature estimate for the site of the species occurrence. A useful adjunct to the distribution map of a tropical montane species is a topographic profile.

This shows the elevation in relation to latitude and indicates whether attention to both temperature and elevation is recommended (e.g. Hall, 1984). Species adapted for prevailing temperatures at high elevations in equatorial latitudes may occur more widely where their ranges approach or transcend the limits of the tropics. On the basis solely of temperature relationships, larger areas will appear suitable toward the latitudinal limits of the range. Species absence in an area reflects the unsuitability of secondary factors or the operation of a barrier.

(c) Linear parameters

The most widespread linear features which determine species distributions are distinctive maritime and riverain conditions. However, a

29

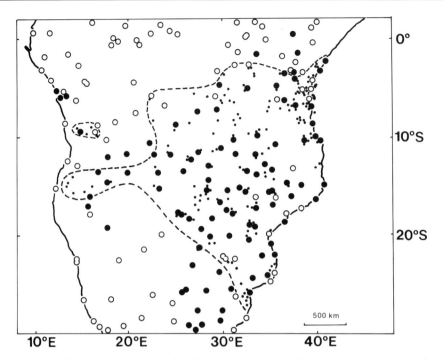

FIGURE 2.3 A species distribution that is related to a point-mapped parameter. The mean number of wet ($\geqslant 50$ mm rainfall) months per annum at meteorological stations in Africa that are <1500 m in elevation and are located between the latitudes 2°N and 30°S are shown together with the range of *Afzelia quanzensis* Welw. *A. quanzensis* is treated as having an essentially continuous distribution within the area delimited by the broken line (one discrete outlier occurs to the west). Dots denote the recorded localities within the species range and circles denote the meteorological station locations. The open circles represent locations with <5 and >7 wet months. The infilled circles represent locations with 5–7 wet months. Refer to Table 2.1 for the details of the tests of association.

TABLE 2.1 The results of a goodness of fit test for *Afzelia quanzensis* Welw. These results show the strength of the association between the natural species range and the number of wet months per annum

Distribution of observations at 187 meteorological stations:

| | Months with $\geqslant 50$ mm mean rainfall | | | |
	≤ 4	5	6–7	>7
Meteorological stations within range of species	7	39	35	16
Meteorological stations outside range of species	25	8	17	40

Statistical analysis of association of different numbers of wet months with the range of species:

	Degrees of freedom	G^2	G^2_{adj}	P
<5 vs >5	1	2.222	—	n.s.
≤ 4 vs 5	1	30.860	30.235	<0.001
6–7 vs >7	1	16.652	16.423	<0.001
Total*	3	49.734	49.044	<0.001

* Partitioned G^2 values sum to the entry in the last row only in the form unadjusted for Type I error.

linear pattern may also arise for species that inhabit the interface between the two dominant tropical vegetation formations of forest and savannah. Identification of such a relationship becomes increasingly difficult with the regression of forest in the face of progressive clearance for other land use. A consideration of the record origination date may reveal locations at the interface on this date, although today's vegetation cover may differ. Conversely, records from within an area where the original cover was of high forest may prove recent in origin. This reflects the spread of a species concomitant with forest fragmentation and the development of new interfaces.

2.5 Conservation significance of the interpreted maps

In the monograph context, the distribution map integrates knowledge of the spontaneous presence of a species over time – old and recent records are both used. The range delimited is assumed to represent areas where conditions potentially favor establishment of the species. Inferences from this association with definable environmental conditions offer guidance to management about where attention to the species is justified and where it is not.

In addition, the map constitutes a valuable framework for systematic consideration of the species' biological and ecological character. Expression of spatial variation in characteristics of interest is possible simply by indicating occurrences with an appropriate series of symbols. It is essential to take account of any systematic geographic variation (e.g. clinal, contrasting infraspecific taxa) if efforts are made to 'domesticate' and 'improve' species. These efforts may involve the enhancement of both product quantity (e.g. stature, fruit size, fruit number) and quality (e.g. tree form, tissue concentration of an extractive – tannin).*

With wide-ranging species, there is often both ecological and morphological variation throughout the species range. This is not apparent from preserved material and seldom evident from collectors' notes. While much taxonomic work takes account of morphological variation, the ecological variation is often neglected and rarely appreciated by taxonomists, beyond generalized comments in species profiles and data sheets. However, critical attention to the ecological literature often exposes much relevant information. With the distribution map as a framework, contrasts in growing conditions or ecological amplitude can be emphasized on the map with symbols or annotated subdivisions.

It is variations concerning product characteristics and ecological amplitude (and tolerance) that the resource sectors consider in developing conservation programs (e.g. Hall, 1992). The distribution map locates populations that offer appealing character in both of these factors. As map labels are assigned to populations in different parts of the range, a foundation is laid for an orderly and thorough conventional improvement process. This is based on the concepts of provenances, provenance trials, and provenance × site matching (Figure 2.4).

Closer consideration of speciation processes and migration routes are encouraged when links between geographic patterns of morphological and ecological variation are revealed. In the light of species' biology (e.g. particularly dispersal agencies), reconstructions on the distribution map of migration routes that take account of paleoenvironments may assist in the understanding of disjunctions and the importance of barriers (Figure 2.1). Where environmental conditions cause migration routes to diverge, any interval wider than the dispersal range that separates populations implies little

---

* When geographical trends are pronounced in such characteristics, the value of the resource will probably not apply throughout the entire range of the species.

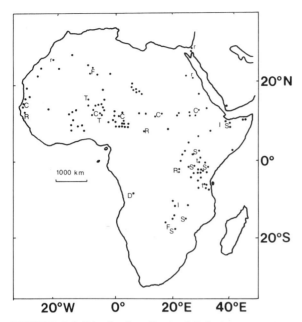

FIGURE 2.4 Distribution (dots) of *Balanites aegyptiaca* (L.) Delile annotated (letters) to provide a basis for developing a conservation strategy. These symbols connote: C, west–east series of populations on vertisolic soils in similar climate for assessment of clinal pattern; D, apparently disjunct population in Angola; F, populations in localities subject to frost; I, location of populations of acknowledged segregate infraspecific taxa; R, populations at high rainfall extremes; r, populations at low rainfall extremes; S, localities of populations associated with congeneric species; T, populations subject to high mean annual temperatures; t, populations subject to relatively low mean annual temperatures.

or no prospect of direct gene exchange. Events subsequent to colonization of currently occupied sites may have created a disjunction by the elimination of populations present on sections of the dispersal route. On other parts of the dispersal routes, where there is less fragmentation, genotypic contrasts (i.e. visible contrasts) are probably less pronounced, even over greater distances.

Barriers and disjunctions for the species that collectively constitute today's vegetation cover often do not coincide. The consequence is that widely ranging species do not have a constant complement of associated species. The idea of associated species is a major part of the ecology of a species, providing insight into the composition of the communities to which it belongs and allowing informed judgement of the role played within these communities. The variations in associates of a species are difficult to convey succinctly and lucidly without the clear geographical framework that is essential to synecological work. By distinguishing geographical subdivisions of the range, and by adopting an appropriate definition of 'associated', a series of related listings (Table 2.2) can be drawn up and ordered. The various relevant scientific information sources contribute the associates for the various subdivisions of the range. Subdivisions can be delimited in any useful manner to ensure that anomalous or invasive circumstances are represented (e.g. Figure 2.5) by river catchments.

2.6 Conclusions

There is a long tradition of distribution mapping. Map resolution varies widely across a spectrum that includes:

(1) scaled-up and regionally adapted versions of the plant geographers' simplistic summaries;
(2) grid-based approaches that are now standard in the conservation sector of industrialized countries; and
(3) attempts to indicate individual sites of occurrence for a particular species.

In the context of monographs of tropical species, the last of these approaches is required but additional information is essential. Early species monographs prepared with the forest sector in mind reveal the limited ecological value of these documents. Not only are they biased toward conventional forestry but the included maps lack interpretive content. In effect, the new generation of monographs necessitates a new generation of distribution

TABLE 2.2 Woody species in different parts of the range of *Acacia seyal* Delile var. *fistula* (Schweinf.) Oliv. that are noteworthy associates of this species

	North–South[a]							
	A	B	C	D	E	F	G	H
Acacia senegal (L.) Willd.	A							
A. polyacantha Willd.	A		C					
Balanites aegyptiaca (L.) Delile	A	B			E			
Acacia drepanolobium Sjóstedt	A		C	D		F		
A. seyal Delile var. *seyal*	A			D		F		
A. sieberana DC.	A			D		F		
Albizia anthelmintica Brongn.					E			
Acacia nigrescens Oliv.							G	
Capparis tomentosa Lam.							G	
Combretum hereroense Schinz							G	
Maerua andradae Wild.							G	
Dichrostachys cinerea (L.) Wight and Arn.							G	H
Acacia kirkii Oliv.								H
A. mellifera (Vahl) Benth. subsp. *detinens* (Burch.) Brenan								H
A. nilotica (L.) Delile subsp. *kraussiana* (Benth.) Brenan								H
Albizia versicolor Oliv.								H
Lonchocarpus capassa Rolfe								H
Peltophorum africanum Sond.								H
Phoenix reclinata Jacq.								H
Ziziphus mucronata Willd.								H

[a] The letters denote: A, Nile catchment; B, Ethiopian Rift Valley catchments; C, Lake Victoria catchment; D, East African Rift Valley catchments; E, East African coastal catchments – Galana/Ruvu systems; F, East African coastal catchments – Wami/Rufiji systems; G, South-eastern African coastal catchments; H, Zambezi catchment.

maps. Where management or conservation measures are needed, time for action is more limited than in the past. Opportunities to learn by trial-and-error and lengthy programs of monitoring are fast receding. Swift responses demand robust baseline information. A modern monograph offers this by its rigorous and comprehensive coverage of all aspects of existing knowledge: authoritative, well-interpreted distributional information is at the core.

2.7 Summary

Attention is drawn to the emergence of numerous tree species as recognized resources which hitherto have received negligible study. A reliable distribution map is a significant prerequisite to augment the collation of knowledge for efficient and orderly interpretation. Information extracted from five sources (herbaria, taxonomic literature, ecological literature, inventories and personal observations) may be represented in a distribution map. Environmental information that is readily available and likely to display patterns of variation related to species distributions includes terrain, climate and soils data. Accessible sources of such environmental data are indicated. Procedures for map production and subsequent interpretation in a management and conservation context are discussed. In map interpretation distinctions are drawn between continuous and fragmented distribution patterns and between environmental information of isoline, mosaic/point and linear types. Mapped distributions of

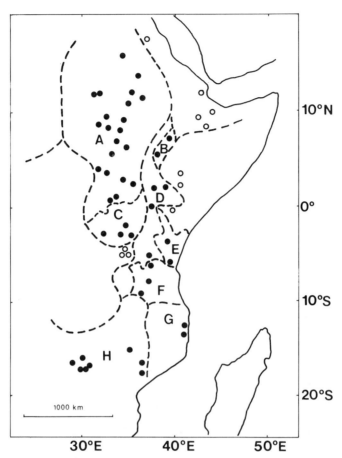

FIGURE 2.5 The shown catchment areas are a basis for a structured summary of the ecological associates of the wide-ranging taxon *Acacia seyal* Delile var. *fistula* (Schweinf.) Oliv. The circles denote the reported localities of occurrence for this species and the broken lines indicate watersheds. The letters identify those catchments (infilled occurrence symbols) for which information is available on associates (see Table 2.2). No information on associates is available for catchments where the occurrence symbols are open. The letters denote: A, Nile catchment; B, Ethiopian Rift Valley catchments; C, Lake Victoria catchment; D, East African Rift Valley catchments; E, East African coastal catchments – Galana/Ruvu systems; F, East African coastal catchments – Wami/Rufiji systems; G, South-eastern African coastal catchments; H, Zambezi catchment.

Acacia seyal vars *fistula* and *seyal*, *Afzelia quanzensis*, *Balanites aegyptiaca* and *Juniperus procera* are used to illustrate the range of variation in these data. Interpretation approaches include the derivation of basic relationships between the species data and the environmental factors, inferences drawn from these relationships, and implications for management and conservation.

References

Bamps, P. (1975) Itinéraire et lieux de récolte de Mildbraed lors de sa première expédition en Afrique centrale (1907–1908). *Bulletin du Jardin Botanique National de Belgique*, 45, 159–79.

Bamps, P. (1982) *Flore d'Afrique Centrale* (Zaire–Rwanda–Burundi). Répertoire des Lieux de Récolte, Jardin Botanique National de Belgique, Meise.

Baumer, M. (1983) *Notes on Trees and Shrubs in Arid and Semi-arid Regions*, FAO, Rome.

Bonkoungou, E.G. (1986) *Etude Monographique du Néré*, CNRST, Ouagadougou.

Bonkoungou, E.G. (1987) *Monographie du Karité Butyrospermum paradoxum (Gaertn. f.) Hepper, Espèce Agroforestière à Usages Multiples*, CNRST, Ouagadougou.

Carlowitz, P.G. von (1986) *Multipurpose Tree and Shrub Seed Directory*, ICRAF, Nairobi.

Dadson, S.F. (1965) *Gazetteer: Federal Republic of Nigeria*, Director of Federal Surveys, Lagos.

Dijkman, M.J. (1950) *Leucaena* – a promising soil erosion control plant. *Economic Botany*, **4**, 337–49.

FAO (1984) *Agroclimatological Data for Africa*, FAO, Rome.

FAO (1985) *Agroclimatological Data for Latin America and the Caribbean*, FAO, Rome.

FAO (1987) *Agroclimatological Data for Asia*, FAO, Rome.

FAO–UNESCO (1977) *Soil Map of the World, 1:5 000 000*, Vol.6, *Africa*, UNESCO, Paris.

Forest Division (1984) *Trees for Village Forestry*, MLNRT, Dar es Salaam.

Friis, I. (1992) *Forests and Forest Trees of Northeast Tropical Africa*, HMSO, London.

Habit, M.A. (1981) *Prosopis Tamarugo: Fodder Tree for Arid Zones*, FAO Plant Production and Protection Paper 25, pp. 1–110.

Hall, J.B. (1984) *Juniperus excelsa* in Africa: a biogeographical study of an Afromontane tree. *Journal of Biogeography*, **11**, 47–61.

Hall, J.B. (1992) Ecology of a key African multipurpose tree species, *Balanites aegyptiaca* (Balanitaceae): the state-of-knowledge. *Forest Ecology and Management*, **50**, 1–30.

Hall, J.B. and Bada, S.O. (1978) The distribution and ecology of obeche (*Triplochiton scleroxylon*). *Journal of Ecology*, **67**, 543–64.

Hepper, F.N. and Neate, F. (1971) Plant collectors in West Africa. *Regnum Vegetabile*, **74**, 1–96.

Letouzey, R. (1968) *Etude Phytogéographique du Cameroun*, Lechevalier, Paris.

Mann, H.S. and Saxena, S.K. (1980) *Khejri (Prosopis cineraria) in the Indian Desert – its Role in Agroforestry*, CAZRI, Jodhpur.

Meteorological Office (1959) *Tables of Temperature, Relative Humidity and Precipitation for the World.* II. *Central and South America, the West Indies and Bermuda*, HMSO, London.

Meteorological Office (1966) *Tables of Temperature, Relative Humidity and Precipitation for the World.* IV. *Asia* (2nd edn), HMSO, London.

Meteorological Office (1983) *Tables of Temperature, Relative Humidity, Precipitation and Sunshine for the World.* IV. *Africa, the Atlantic Ocean South of 35°N and the Indian Ocean* (2nd edn), HMSO, London.

National Research Council (1975) *Underexploited Tropical Plants with Promising Economic Value*, National Academy of Sciences, Washington.

National Research Council (1977) *Leucaena: Promising Forage and Tree Crop for the Tropics*, National Academy of Sciences, Washington.

National Research Council (1980) *Firewood Crops*, National Academy of Sciences, Washington.

National Research Council (1983a) *Firewood Crops*, Vol. 2, National Academy of Sciences, Washington.

National Research Council (1983b) *Calliandra: a Versatile Small Tree for the Humid Tropics*, National Academy of Sciences, Washington.

National Research Council (1983c) *Casuarinas: Nitrogen-Fixing Trees for Adverse Sites*, National Academy of Sciences, Washington.

Oakes, A.J. (1968) *Leucaena leucocephala*: description, culture, utilization. *Advancing Frontiers of Plant Sciences*, **20**, 1–114.

Panday, K. (1982) *Fodder Trees and Tree Fodder in Nepal*, Swiss Development Co-operation, Berne.

Polhill, D. (1988) *Flora of Tropical East Africa: Index of Collecting Localities*, Royal Botanic Gardens, Kew.

Reilly, P.M. (1976) Sabah, Malaysia. *Ministry of Overseas Development Land Resource Bibliography*, **8**, 1–94.

Teel, W. (1984) *A Pocket Directory of Trees and Shrubs in Kenya*, KENGO, Nairobi.

White, F. (1983) *The Vegetation of Africa*, UNESCO, Paris.

Wickens, G.E. (1972) Dr G. Schweinfurth's journeys in the Sudan. *Kew Bulletin*, **27**, 129–46.

Withington, D., Glover, N. and Brewbaker, J.L. (1987) *Gliricidia sepium (Jacq.) Walp.: Management and Improvement*, NFTA, Honolulu.

Chapter three

Mapping the elements of biodiversity: The rare species of Madagascar

Ronald I. Miller and Jennifer H. Allen

3.1 Introduction

3.1.1 A perspective

Preservation of areas of high rare species richness will be a critical consideration in the preservation of biological diversity in the twenty-first century (Wilson, 1988; May, 1992). Techniques by which to identify these areas more precisely are urgently needed (Sullivan, 1992). Mapping these locations can assist governments and development agencies in integrating biodiversity conservation into their programs.

The pilot study described in this chapter focused on the development of a methodology to document and graphically display the location of areas rich in threatened or endangered species in Madagascar. While these rare species constitute only a subset of biological diversity in general, changes in patterns of endangered species richness may serve as an indicator of broader trends in species distribution (Cody, 1986).

3.1.2 The island of Madagascar

Madagascar was chosen as the focus for this study because of the importance of its plant and animal life, the pressures for development and the availability of key data. Madagascar has been identified as one of seven 'mega-

Mapping the Diversity of Nature. Edited by Ronald I. Miller.
Published in 1994 by Chapman & Hall, London. ISBN 0 412 45510 2.

diversity countries' by the World Wide Fund for Nature (WWN) and the International Union for the Conservation of Nature and Natural Resources (IUCN). These countries contain an endowment of plant and animal life so rich and diverse that they constitute biological treasures of value to the entire planet. More than three-quarters of the native species of Madagascar occur nowhere else in the world, and little is known about the characteristics and behavior of many of these species. These characteristics have made the protection of Madagascar's remaining habitat a priority for the international scientific and conservation community.

Adding urgency to the need to document the location of Madagascar's species is the accelerating rate of environmental destruction on the island. Madagascar's landscape and natural resources have suffered from the spread of settlement and agricultural development (Olson, 1988), and in recent years a number of factors have contributed to accelerated degradation across the island. The nation's population has doubled over the past 20 to 25 years, fueling land clearance for agricultural production. While forests once covered most of the island, only one-tenth of the land areas remain forested, a factor that contributes to high levels of erosion in many areas (Green and Sussman, 1990). In addition, poor security of land tenure provides little incentive for farmers to invest in the land, particularly with regard to soil conservation measures. Adding to the island's vulnerability is its rugged topography and the frequent torrential rains, which further exacerbate erosion rates.

Environmental degradation threatens Madagascar's economic development as well as its plant and animal species. The World Bank estimates the cost of environmental degradation in Madagascar at between $100 million and $300 million per year – from 5 to 15 % of the country's gross national product (Young, 1989). Wise management of the country's natural resources is therefore a concern of development planners as well as of the conservation community.

3.1.3 The objectives of this study

This chapter presents a broad but potentially valuable method for documenting and graphically displaying the location of rare species richness in a developing country. By choosing the most reliable data available and enlisting the assistance of experts with first-hand knowledge of the area to refine the species ranges derived from these data, credible estimates of the present location of plant and animal species can be made. Presenting these estimates in map form makes the data accessible to a wide variety of users. Development personnel in particular are often constrained by lack of information about environmental factors; even when data do exist, they are often in a format which is difficult to use in project planning. This method uses data available in the literature and presents it in a graphical format which can be used in development project planning.

One of the main objectives of this approach is to integrate data from a variety of sources and to present these data in a standardized form. Scientific data are collected in a wide variety of formats which are often difficult to adapt to alternative applications. Field notes are especially difficult to use for environmental assessments, as they are often anecdotal and geographically imprecise. The approach presented in this study attempts to address these obstacles by using expert review to improve the reliability of the original data, and by presenting the data in a readily accessible graphical format.

3.2 The approach

3.2.1 Introduction

Scientific data are initially collected for many different reasons. The formats of these original

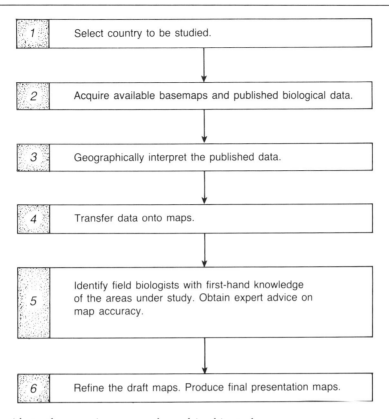

FIGURE 3.1 A guide to the mapping approach used in this study.

data are often not easily adaptable to alternative applications. Species data collected as field notes are especially difficult to use directly for environmental assessment. The presentation of these data in map format significantly enhances their usefulness in environmental planning and management.

Maps can serve to accurately present the spatial status of biological data. Well-produced maps can easily be interpreted by a wide variety of experts, as well as by the general public. The broad-scale perspective of the maps can help improve inventories of natural resources; impact assessments of development projects; and project planning, design and management. The species mapping approach used here descended from past studies of rare plants in the southern Appalachian mountains of the

United States (Miller, 1986) and rare birds in Tanzania (Miller et al., 1989).

3.2.2 An overview of the methodological approach

The pilot study involved a six-step process, which is summarized in Figure 3.1. The first step was the selection of the appropriate developing country. Next the published data on rare and endangered species were collected, together with available maps of the country. The third step was to interpret the published data, primarily derived from biologists' field notes, for mapping.

Once the data had been evaluated and geographically referenced, the ranges indicated in the notes were drawn onto the composite base maps. The fifth step was to obtain expert advice regarding the range estimates from field biologists familiar with the species of concern. This step incorporates the 'Delphi Method' from physical geography (Luscombe, 1986), and this permitted the production of reasonably accurate maps with the available data. Finally, the maps were revised based upon the expert opinions, and final maps were produced. The following sections discuss each of these steps in greater detail.

3.2.3 Country selection

The country selected for this study needed to fill the conditions of both the international development and the international conservation communities. Three primary criteria were used to select the country for study:

1. The need for immediate conservation action in the country. Countries under consideration for this study were vulnerable to intensive development and/or potentially rich in rare and threatened species.
2. The importance of conservation to the country and the quality of the available data.
3. The prominence of the country in the World Bank's work program. The World Bank was the voice of the development community during this study.

Madagascar was chosen as the focus for this study because of the importance of its plant and animal life, the pressures for development, the availability of key data, and the need to promote sustainable development on the island.

3.2.4 Basemap production

Since maps were not available that included all of the geographical features referred to in the field notes, composite maps were developed using the GIS (i.e. PC ARC/INFO). The basemaps for Madagascar were formulated to present the location of national boundaries, selected rivers, lakes and cities. The composite maps included topographic lines; mountain ranges and specific peaks referenced in the field notes; rivers referred to in the field notes; major cities and towns; and a number of other reference points required for range estimation. The geographic feature locations were taken from a series of three 1:1 000 000 Operational Navigation Charts (ONC) that cover the northern, central and southern areas of Madagascar.

3.2.5 Species selection

Biologists' field notes are the primary source of species observation data for Madagascar. These observations have been recorded over the past 150 years and they form the basis of published and unpublished species distribution records in Madagascar. This pilot study relied primarily on data published in *Madagascar: An Environmental Profile* (IUCN/UNEP/WWF, 1987). For some of the rare lemurs, published range maps (Tattersall, 1982) were also consulted.

The threatened species considered in this study are listed in Table 3.1. The birds, butterflies, lemurs and reptiles were selected as key species groups on the island, since they fall under one of the four IUCN threat categories listed in Table 3.2, but species in the 'insufficiently known' threat category were excluded from consideration. The Madagascar species field data used for this study are listed in Table 3.3. The screening techniques used to cull and interpret these data for map production are presented the following section.

3.2.6 Documenting the locations of place-names and features

Several interpretive approaches were taken to cope with the variability in the data — for example, species observations often refer to locations (e.g. cities, mountain ranges, rivers, etc.) with place-names that have changed over

TABLE 3.1 The threatened species of Madagascar examined during this project

LEMURS
Allocebus trichotis (Hairy-eared dwarf lemur)
  Endangered species, endemic to Madagascar, and considered to be the rarest of all the lemurs. Known to occur in the eastern tropical forest.
Indri indri (Indri)
  The largest living lemur, an endangered species, restricted to parts of the northeastern tropical forests of Madagascar.
Propithecus diadema (Didemed sifaka)
  Vulnerable species, endemic to eastern Madagascar; this lemur occurs naturally at low population densities and is severely threatened due to habitat loss from deforestation.
Daubentonia madagascariensis (Aye-aye)
  Endangered species, endemic to Madagascar; rare due to the destruction of the forest habitat.
Hapalemur simus (Broad-nosed gentle lemur)
  Endangered species, endemic to Madagascar; known only from the southeast region of the island. Found along river banks where the Giant Bamboo species occurs.
Lemur macaco macaco (Black lemur)
  Vulnerable species, endemic to northwest Madagascar, found in the humid forests and on the coastal islands.
Lemur macaco flavifrons (Sciater's lemur)
  Endangered species, endemic to a coastal forest strip in the northwest of Madagascar.
Lemur mongoz (Mongoose lemur)
  Vulnerable species, occurs in the deciduous forests of northwest Madagascar, and also occurs in the Republic of the Comoros.
Lemur rubriventer (Red-bellied lemur)
  Threatened species, endemic to the eastern forests of Madagascar.
Varecia variegata (Ruffed lemur)
  Threatened species, endemic to the eastern humid forests of Madagascar.

REPTILES
Geochelone radiata (Radiated tortoise or Sokake)
  Vulnerable, endemic, terrestrial species restricted to dry forest.
Geochelone yniphora (Angonoka)
  Endangered, endemic, terrestrial species restricted to an area around Baly Bay. In imminent danger of extinction.
Pyxis planicauda (Madagascar flat-tailed tortoise of Kapidolo)
  Restricted, endemic, terrestrial species threatened by agricultural encroachment and habitat destruction.
Pyxis arachnoides (Madagascar spider tortoise or Tsakafy, Kapila)
  Restricted, endemic, small terrestrial tortoise species found in xeric thorn-bush scrub of coastal regions.
Erymnochelys madagascariensis (Madagascar sideneck turtle or Rere)
  Restricted, endemic, aquatic turtle species, found in extensive aquatic habitats on the west and northwest of the island.

BUTTERFLIES
Papilio grosesmithi
  Rare, endemic, butterfly species known from the deciduous forests of western Madagascar.
Papilio morondavana
  Vulnerable, endemic, attractive swallowtail butterfly species found in the deciduous forests in western Madagascar. Seriously threatened by habitat destruction.
Papilio mangoura
  Rare, endemic, butterfly species restricted to some of the eastern rainforests.

time. When Madagascar became independent from France in 1972, many French location names were replaced by Malagasy names, and by comparing maps published over the past 40 years it was possible to correlate the names used in the field notes with current place-names.

Landscape features provide the biologists

TABLE 3.2 The IUCN rarity category definitions used to select the threatened species of Madagascar considered in this project

Endangered (E): Taxa in danger of extinction and whose survival is unlikely if the causal factors continue to operate.

Vulnerable (V): Taxa likely to move into the endangered category in the near future if the causal factors continue to operate.

Rare (R): Taxa with small global populations, not presently classified as endangered or vulnerable, that are at risk.

Indeterminate (I): Taxa known to be endangered, vulnerable or rare but with insufficient information to categorize appropriately at present.

who proof and edit species maps with reference features that improve the precision of the mapped species distribution ranges. In some cases, species ranges are described in reference to landscape features. The locations of these landscape features then need to be approximated from pre-existing maps or other data sources. Some of the landscape features that biologists referred to in their species observations were not clearly demarcated on available maps. The expert reviewers' general knowledge of the island was helpful in approximating the location of such features.

3.2.7 Interpretation of the field observations

Field notes present difficulties for precise mapping. Until very recently, the clear documentation of geographic location information was usually not a foremost concern of a biologist recording observations in the field. Consequently, deriving site information from field notes can be a difficult and arduous task, and the level of accuracy obtained from interpretations of these data is questionable. If species observations are to be mapped with any degree of confidence, the development of a standardized methodology for interpreting field notes is

a necessary and critical step in the mapping process.

Some of the challenges presented in the interpretation of field notes can be illustrated by an example. The following excerpt, from the field notes for the Madagascar flat-tailed tortoise, is reprinted here from *Madagascar: An Environmental Profile* (IUCN/UNEP/WWF, 1987):

> An endemic Madagascar species. Apparently restricted to the Andranomena forest, an area of approximately 100 square kilometers situated 20 kilometers northeast of Morondava on the central-west coast of Madagascar. Records outside this area are unconfirmed, the species may occur as far north as Maintirano but no specimens have been found in apparently suitable forests around the Andranomena area.

The boundary of the forest referred to in this entry was not indicated on available maps, so an approximate boundary was estimated. The notes were condensed to the following entry in an appendix:

> Andranomena Forest. Restricted to this forest (100 km²) on the central-west coast of the island. Endemic.

A precise, bounded location for this forest was probably not accessible from available maps. Thus a polygon was drawn in this area of the island to approximate this location.

Information from unsubstantiated secondary sources and data from older museum specimens was not considered in this study to secure the highest level of data validity possible. Therefore, the study did not map most reported records prior to 1930 unless modern sightings confirmed these records. A few species were known to exist in Madagascar but a current range could not be confirmed from recent sitings. For these few species, unconfirmed observations recorded prior to 1930 were used

TABLE 3.3 A compilation of distribution records for selected threatened species in Madagascar belonging to the taxa: M = Mammalia (lemurs); A = Aves (birds); L = Lepidoptera (butterflies); and R = Reptilia (reptiles). These data were interpreted and compiled from distributional records published in *Madagascar: An Environmental Profile* (IUCN/UNEP/WWF, 1987). The listed distributional ranges include only confirmed records collected since 1930. Nonendemic chelonian species are not included. The species numbers (#) are assigned solely for reference purposes.

Taxon	#	Species	Distribution data	Geographical coordinates	Category
M	1	*Allocebus trichotis* (Hairy-eared dwarf lemur)	Andranomahitsy forest – to west of Mananara on the east coast.	16°00′S, 50°00′E	E
M	2	*Indri indri* (Indri)	Rainforest along the northeastern escarpment between the latitudes of Sambava and Mahanoro, not on Masaola peninsula.	14°20′S, 50°10′E 17°50′S, 48°50′E	E
M	3	*Propithecus diadema* (Didemed sifaka)	Forests in the following areas (five subspecies inclusive): – between the Mangoro river and the latitude of Maroantsetra (not in immediate vicinity of town) – north of Maroantsetra to the Andapa basin and the Marojejy massif – south of the Mangoro river to the latitude of Manakara – strip of forest between latitudes of Fandriana and Vondrozo, near Fianarantsoa – northeast of the Andrafiamena mountains, south and east of Anivorano Nord Type site: dry forest of Analamera.	See Tattersall (1982, p. 100) for reference coordinate data	V
M	4	*Daubentonia madagascariensis*	Lowland forest near Mananara south of Maroantsetra; mid-altitude forest around Perinet; Nosy Mangabe in Antongil Bay; Anosyenne hills northwest of Taolanaro, in the Andohahela Natural Reserve; Montagne d'Ambre to Ankobakabaka near Befandriana Nord; Ampasimena – at the northern tip of the Ampasindava peninsula.	16°10′S, 49°46′E 18°56′S, 48°24′E 15°45′S, 49°50′E 24°40′S, 46°40′E 12°30′S, 49°20′E 15°10′S, 48°30′E 13°50′S, 48°00′E	E
M	5	*Hapalemur simus* (Broad-nosed gentle lemur)	Southeast forests: Ranomafana; east at Kianjavato near Mananjary.	21°16′S, 47°28′E 21°30′S, 47°50′E	E
M	6	*Lemur macaco macaco* (Black lemur)	Nosy Be and Nosy Komba islands; also extends from Anivorano Nord to Befandriana to just south of Maromandia (on the coast); includes Tsaratanana massif and the Ampasindava peninsula.	13°50′S, 48°00′E 12°45′S, 49°10′E 15°15′S, 48°50′E 14°15′S, 48°25′E 13°50′S, 48°30′E 13°45′S, 48°00′E	V
M	7	*Lemur macaco flavifrons* (Sciater's lemur)	Along northwest coast in a forest strip between Maromandia and Befotaka about 100 km north of Ampasindava Bay.	14°15′S, 48°25′E 14°50′S, 48°00′E	E

TABLE 3.3 *continued*

Taxon	#	Species	Distribution data	Geographical coordinates	Category
M	8	*Lemur mongoz* (Mongoose lemur)	Northwest deciduous forests:		V
			Lake Kinkony	16°20′S, 45°50′E	
			to the south of Mitsinjo,	15°55′S, 45°45′E	
			to the west of the river Mahavavy,	16°03′S, 45°55′E	
			but not known from the Tsingy de Namoroka Reserve (20 miles south of Soalala).	16°10′S, 45°15′E	
			To the east and west of the Betsiboka river near Ambato-Boeni,	16°30′E, 46°40′S	
			north to Bay of Narinda.		
M	9	*Lemur rubriventer* (Red-bellied lemur)	Throughout the mid-altitude eastern forest:		V
			northern limit – Tsaratanana massif;	13°50′S, 48°30′E	
			southern limit – Ivohibe at southern end of Andringitra massif.	22°30′S, 47°00′E	
			Zahamena Reserve	17°30′S, 49°00′E	
			Perinet	18°55′S, 48°20′E	
			Fianarantson, around Ranomafana;	21°16′S, 47°28′E	
			also at Kianjavato, due east of above	21°10′S, 47°50′E	
M	10	*Varecia variegata* (Ruffed lemur)	Eastern forests: north and west of Maroantsetra,	15°23′S, 49°45′E	I
			to south of Farafangana,	23°00′S, 47°50′E	
			but north of Mananara river;		
			Nosy Mangabe;	15°30′S, 49°40′E	
			Masoala peninsula to the east of the Antainambalana river.	15°45′S, 50°15′E	
R	11	*Geochelone radiata* (Radiated tortoise or Sokake)	Didierea forest:		
			restricted to this forest – a narrow arc across southern Madagascar – from Amboasary in the SE to	25°10′S, 46°10′E	V
			Morombe in the SW, endemic.	21°50′S, 43:20′E	
R	12	*Geochelone yniphora* (Angonoka)	Restricted to three forest islands in the vicinity of Baly Bay (area of 60.25 km²) NW, endemic.	16°00′S, 45°20′E	E
R	13	*Pyxis planicauda* (Madagascar flat-tailed tortoise or Kapidolo)	Andranomena forest: restricted to this forest (100 km²) on the central-west coast of the island, endemic.	20°15′E, 44°30′S	I
R	14	*Pyxis arachnoides* (Madagascar spider tortoise or Tsakafy, Kapila)	South and southwest region near the coast – from 10 to 50 km inland:		
			Morombe in the north	21°43′S, 43°20′E	
			Amboasary (near Ft Dauphin) in the south, endemic.	25°10′S, 46°10′E	
R	15	*Erymnochelys madagascariensis* (Madagascar sideneck turtle or Rere)	Aquatic habitats W–NW of island:		I
			from Mangoky river	21°43′S, 43°45′E	
			and Lake Ihotry	21°59′S, 43°36′E	
			near Marombe in SW	21°43′S, 43°20′E	
			north to the Sambirano basin	13°55′S, 48°30′S	
			west of the Massif de Tsaratanana, endemic.	14°00′S, 49°00′E	

TABLE 3.3 *continued*

Taxon	#	Species	Distribution data	Geographical coordinates	Category
L	16	*Papilio grosesmithi*	Western Madagascar: from Mahajanga (Majunga) in the north to Sakaraha,	15°50′S, 46°05′E	R
			Toliara and	23°10′S, 44°30′E	
			the Lambomakondro forest in the south, endemic.	23°30′S, 43°50′E	
L	17	*Papilio morondavana*	Western forests: region around towns of Morondava	20°15′S, 44°20′E	R
			and Mahabo	20°22′S, 45°00′E	R
			north toward Mahajanga	15°50′S, 46°05′E	
			and Ambato-Boeny,	16°30′S, 45°48′E	
			and south to Andranovory		
			and Toliara.	23°30′S, 43°50′E	
L	18	*Papilio mangoura*	Eastern Madagascar, from Maroantsetra in the north	15°30′S, 49°45′E	R
			to Taolanaro in the south.	25°05′S, 47°00′E	
A	19	*Tachybaptus rufolavatus* (Alaotra grebe)	Lake Alaotra and adjacent marshes (ca. 700 m²) – the only known breeding site.	17°30′S, 48°30′E	E
A	20	*Anas bernieri* (Madagascar teal)	Western coast:		V
			S – Lake Ihotry, southeast of Morombe;	21°59′S, 43°36′E	
			N – Montagne d'Ambre;	12°30′S, 49°20′E	
			NW – Ambilobe, to north of Morombe;	21°45′S, 43°15′E	
			W – Antsalova region, especially Lake Bemamba;	18°30′S, 44°45′E	
			Lake Kinkony;	16°20′S, 45°50′E	
			Lake Bemamba.	18°55′S, 44°20′E	
A	21	*Aythya innotata* (Madagascar pochard)	North central plateau: lakes and pools; Lake Alaotra, Lake Ambohibao, Ambadirato – The Andilamena region (70 km N of Lake Alaotra); Betsileo country.	17°30′S, 48°30′E	E
A	22	*Haliaeetus vociferoides* (Madagascar fish eagle)	West coast: from Morondava to Diego Suarez;	12°20′–20°S, 44°–49°20′E	E
			Nosy Be;	13°20′S, 48°20′E	
			Lake Kinkony region: Mahajanga;	15°40′S, 46°30′E	
			Ambararatabe		
			Triangle: Soalala,	16°10′S, 45°25′E	
			Namakia,	15°50′S, 45°48′E	
			Lake Kinkony,	16°20′S, 45°50′E	
			Antsalova region;	18°30′S, 44°45′E	
			Lake Masama: Manambolo river; Rectangle of lakes and marshes between:		
			Antsalova;	18°30′S, 44°45′E	
			Bekopaka;	19°10′S, 44°50′E	
			Sea;	19°00′S, 44°18′E	
			Maintirano.		

TABLE 3.3 *continued*

Taxon	#	Species	Distribution data	Geographical coordinates	Category
A	23	*Eutriorchis astur* (Madagascar serpent eagle)	Known from only eight specimens, all collected more than 50 years ago. Selected records from past sightings are included here. Eastern rainforests; Sihanaka forest. Analamazoatra near Perinet (also near Rogez); Maroantsetra; Bevato – sea level (40 km NW of Maroantsetra); Ambolomirahavavy (NE), 600 m elevation (name cannot be traced); Marojejy Reserve	18°50′S, 48°35′E 18°50′S, 48°30′E 18°55′S, 48°10′E — 14°18′S, 49°33′E	E
A	24	*Mesitornis variegata* (White-breasted mesite)	NE of Morondava, i.e. 10 km SW of Marofandilia, 9 km S and 15 km N–NW of Beroboka, 3 km S of Ampamanrika lake. Ankarafantsika forest; Ankarana Cliffs N – (25 km SW of Tsarakibany).	20°15′S, 44°20′E 20°10′S, 44°30′E 20°00′S, 44°40′E 16°20′S, 46°50′E 12°55′S, 49°10′E	R
A	25	*Monias benschi* (Subdesert mesite)	70 km coastal strip between Mangoky and Fierenana rivers – SW; Fativolo – limit of inland range.	21°40′S, 43°45′E 23°30′S, 44°10′E 23°02′S, 44°10′E	R
A	26	*Sarothrura watersi* (Slender-billed flufftail)	Southeast Betsileo (i.e. south-central Madagascar); Analamazoatra near Perinet; near Andapa (NE) at 1800 m elevation; distribution of this species may be determined by the distribution of rainforest	14°30′–18°55′S, 48°55′–49°40′E 18°50′S, 48°30′E	I
A	27	*Charadrius thoracicus* (Madagascar plover)	Coastal SW Madagascar between Morondava and Androka; Morombe/Mangoky river delta area; between Morombe and Lake Tsimanampetsotsa	20°15′S, 44°20′E 21°43′S, 43°20′–43°45′E 21°43′S, 43°20′E	R
A	28	*Tyto soumagnei* (Madagascar red owl)	Eastern region – in a circle whose diameter runs between Toamasina and Antananarivo.	18–19°S, 47–49°E 18°10′S, 49°20′E 18°50′S, 47°40′E	I
A	29	*Brachypteracias squamiger* (Scaly ground-roller)	Throughout the eastern rainforests of Madagascar; deep rainforest in the center and northeast of Madagascar: Marojejy Maroantsetra Masoala peninsula Perinet Analamazoatra Rogez.	 14°18′S, 49°33′E 15°25′S, 49°35′E 15°45′S, 50°15′E 18°30′S, 48°20′E 18°50′S, 48°30′E 18°50′S, 48°35′S	R

TABLE 3.3 *continued*

Taxon	#	Species	Distribution data	Geographical coordinates	Category
A	30	*Brachypteracias leptosomus* (Short-legged ground roller)	Occurs in two discrete general areas of Madagascar: Northeast – Marojejy to Maroantsetra; central east – Sihanaka forest	14°18′S, 49°33′E 15°25′S, 49°35′E	R
A	31	*Atelornis crossleyi* (Rufous-headed ground roller)	Northeast: Tsaratanana massif, Marojejy Reserve, Andapa; Central–east: circle with a diameter from Anatananarivo to Toamasina; Vondrozo region	13°50′S, 48°30′E 14°18′S, 49°33′E 14°30′S, 49°45′E 18°50′S, 47°40′E 18°10′S, 49°20′E	R
A	32	*Uratelornis chimaera* (Long-tailed ground roller)	SW coastal strip: between Mangoky river; and the Fiherenana river – up to 80 m elevation.	21°–22°S, 43°30′–44°00′E 21°45′S, 44°30′E 23°30′S, 44°10′E	R
A	33	*Neodrepanis hypoxantha* (Yellow bellied sunbird-asity)	SE of Imerina plateau in Sihanaka forest; forest near Perinet – Analamazoatra Special Reserve – 100 m higher than reserve	16–18°S, 48–49°E	I
A	34	*Phyllastrephus apperti* (Appert's greenbul)	Forest 40 km SE of Ankazoabo in SW Madagascar; also SE in nearby Zombitsy forest (= Mangona = Vihibasia = Jarindrano) – E of Sakaraha and east and south of Andronolava	23°10′S, 44°30′E	R
A	35	*Phyllastrephus tenebrosus* (Dusky greenbul)	Perinet; Perinet – Analamazoatra Special Reserve	18°30′S, 48°20′E 18°28′S, 48°28′E	R
A	36	*Phyllastrephus cinereiceps* (Grey-crowned greenbul)	Tsaratanana massif; near Didy on the western edge of the Sihanaka forest.	14°00′S, 49°00′E 18°10′S, 48°30′E	R
A	37	*Xenopirostris damii* (Van Dam's vanga)	Ankarafantsika plateau, SE of Mahajanga	16°00′S, 47°00′E	R

The following four threatened bird species were not included in this listing: *Xenopirostris polleni, Monticola bensoni, Crossleyia xanthrophrys, Newtonia fanovanae*.

to produce estimates of potential distribution patterns for the maps.

In summary, the data-screening process excluded species observations of doubtful authenticity; those without confirmed observations in the past 25 years; and those with an IUCN rarity classification of 'insufficiently known'. Table 3.4 lists the range estimates upon which the species distributions patterns presented in Colour Plates 1, 2 and 3 are based.

3.3 Species distribution maps

For the production of the presentation maps, the local ranges around the observation records

TABLE 3.4 The following location and area estimations were required to map the species field data. In all cases, the tropical forest band along the eastern coast was approximated (i.e. an estimated distance between the tropical forest band and the eastern coastline of Madagascar). The observed point locations were not independently mapped when they fell within a polygon representing the species range. Circles were drawn to estimate areas indicated by city or village names. The species numbers (#) refer to the species data listed in Table 3.3

Species #	Required range estimations
1	A small circle was used to estimate the position at 16°10'S and 49°30'E.
2	The forest range was mapped to follow along the eastern ridge of the mountains between the latitudes of Sambava and Mahanoro to the coastline, avoiding the Masoala peninsula. The extent of the eastern mountain escarpment was unknown at the time of mapping.
3	A map from Tattersall (1982, p. 100) was used to estimate the ranges. Anivorano Nord was estimated at 12°45'S and 49°10'E.
4	(a) Circles were used to estimate mid-altitude locations around Perinet.
	(b) A circle estimate was used for the position of Ampasimena at the tip of the Ampasindava peninsula.
	(c) A long thin corridor was drawn from M. d'Ambre to Befandriana Nord.
	(d) The Anosyenne hills were estimated with a circle.
5	Circles were drawn around Ranomafana and Kianjavato.
7	A corridor was estimated (including Befotaka city south of Maromandia) almost up to the Bay of Befotaka.
8	The area from Mitsinjo to the Mahavary was estimated west to the reserve but not including this reserve. A corridor was drawn along the river through Ambato-Boeni and then along the forest corridor to the Bay of Narinda.
16	Lambomakondro forest was not able to be located. The boundary was drawn between Toliara and Sakaraha.
17	A strip on the western coast was estimated; the western forest was 'estimated' from the 1959 map.
21	Lakes Ambohibao and Ambadirato were not located on the available maps.
22	The Manambolomaky river was equated with the Manambolo river.
23	A circle was drawn using the IUCN/UNEP/WWF (1987) coordinates for the Sihanaka forest.
24	A circle was drawn around Marofandilia. Lake Ampamanrika was not located on the available maps. Beroboka was equated with Beroboka Nord.
27	Lake Tsimanampetsotsa was not located on the available maps.
28	Toamasina and Anatananarivo were used as points on the perimeter of a circle.
30	Rogez was not located on the available maps.
34	A polygon was drawn at the coordinates of the Analamazoatra Special Reserve.

were estimated from the compiled species observations data (Table 3.3). A variety of required range and point estimations were used to produce the final maps (Table 3.4). Polygons were drawn manually to approximate the species ranges identified from the published data.

Dr Ian Tattersall at the American Museum of Natural History in New York kindly provided guidance for the lemur maps; Dr Aaron Bauer at Villanova University in Villanova, Pennsylvania, provided advice regarding the reptile maps; and Professor R. Paulian at Sainte Foy la Grande in France reviewed the butterfly maps.

Dr Olivier Langrand at the World Wide Fund for Nature in Antananarivo, Madagascar, kindly provided a review of the bird distribution data (however, bird range maps were not produced during this study). The draft map sheets with the expert revisions were then used to produce the final presentation maps in Plates 1–3.

The map sheets, with the revisions from the specialists, were then used to produce the final maps. As a final validation, place-names on the draft map sheets were referenced to a short list of current Malagasy place-names maintained at the World Bank.

3.4 Evaluation and recommendations

In future projects of this type, care should be taken to maximize the number and types of features on the base map. This will increase the utility of the maps if they are used to plan and/ or appraise projects within an institution. The features detail can be achieved either by creating a new composite map from available data or by identifying an appropriate existing map. The species distribution ranges should be mapped directly onto the base map copies that are sent to the biologists for review. The revised maps received from the biologists are then ready for final map production.

The final presentation maps produced in this project are useful for the early stages of the planning process within a country. For the more detailed work that is part of the later stages, the tabulated species data in Table 3.3 provides a more precise guide for measuring the impacts of projects on rare species.

Maps will play an increasingly important role in development planning as procedures are refined to map more complex features such as habitats and ecosystems, but it is always important to remember that maps are only as good as the data and interpretation used to produce them. In this study, rigorous steps were necessary to ensure the reliability and uniformity of data and their interpretation.

An important problem arises in updating field observations and other data used to compile the maps. Data can become dated quickly in many developing countries which are undergoing rapid change. In order to address this problem, development and conservation institutions always need to consider the importance of a systematic program to update information reliably. This updating capability will be essential to sustain any modern mapping and database systems.

GIS capabilities provide many advantages in the areas of environmental/conservation research and planning. In this study, the GIS demonstrated evident long-term advantages in this study for the establishment and maintenance of maps and data regarding species distribution patterns. These benefits include:

- the ability to regularly edit and update species distribution maps
- the ability to produce hardcopy, updated maps on a regular basis
- the practical and scientific values inherent in the components of a computerized database.

3.5 Future applications

Biodiversity maps and their accompanying data have widespread applications for environmental planning and management. They can be an important source of information for Environmental Action Plans. These maps can readily be used to evaluate the impact of planned development projects.

Most broadly, the maps and their accompanying data can alert people to the need to protect vulnerable or sensitive areas. For conservation planning, maps can be used to ensure that proposed park boundaries will encompass the location of important biodiversity resources. The overlay of species distribution maps onto protected areas can be used today (e.g. Miller and White, 1986) to confirm that protected area boundaries are fulfilling their promise.

Taking environmental considerations into account in designing projects in developing countries has often been interpreted as a threat to national autonomy. It is therefore important that the integration of environmental concerns such as the preservation of biodiversity be conducted with the collaboration of the involved countries. The mapping of species, habitat and natural resource patterns in relation to development proposals can serve as a tool for visualizing future impacts.

3.6 Summary

This chapter presents the results of a pilot study conducted at the World Bank in 1988–9 and

49

involved the mapping of rare species distribution patterns in Madagascar. The mapping procedures presented here were derived from an approach first developed for the mapping of rare bird species in Tanzania (Miller *et al.*, 1989).

Maps showing the diversity and abundance of plant and animal species provide a valuable tool for environmental planning and management. Such maps help to pinpoint where species are located so that the species and their habitats can both be better protected and managed. These maps can serve generalists and experts alike, as their graphical format makes them readily accessible to economists, development planners, government officials, experts in other disciplines and the general public.

Using Madagascar as the case study, this study attempted to develop methods to refine species data derived from biologists' field notes. Based on the interpretation of the field notes, boundaries delimiting the ranges where species are most likely to be found were drawn on base maps of the country. The species location data were successively refined through consultations with experts possessing first-hand knowledge of the area and the species under study. Such a consultative technique may prove valuable in other cases in which published data are imprecise and of uncertain validity.

Such maps have a wide range of applications. They can help in the development of environmental action plans, in improving the design of national parks and protected areas, and in ensuring that existing parks have an adequate geographic coverage to protect a country's critical habitats. The maps can assist in evaluating the environmental impact of development projects and in improving project design and management. They can also be used to educate the general public to the need to protect environmentally sensitive areas with important biodiversity components. One of the main

objectives of this project was to develop a methodology to make such maps as reliable as possible.

Acknowledgements

We would like to thank the ENVOS Division of the Environment Department of the World Bank for providing the support and facilities which enabled this study to be conducted. Unit Director Dr D. Jane Pratt provided particular support during the completion of this project. We would also like to thank the personnel in the Center for Earth Resources Analysis (CERA) in ENVOS for their guidance and advice during the course of this study.* The maps depicting the rare species in Madagascar were produced in collaboration with staff in the World Bank's Cartographic Division.

The authenticity of the species maps was significantly enhanced by Dr Ian Tattersall, Dr Aaron Bauer, and Professor R. Paulian.

References

Cody, M.L. (1986) Diversity, rarity, and conservation in Mediterranean-climate regions, in *Conservation Biology: The Science of Scarcity and Diversity* (ed. M.E. Soule), Sinauer Associates, pp. 122–52.

Feoli, E. and Orloci L., (1985) Species dispersion profiles of anthropogenic grasslands in the Italian eastern pre-Alps. *Vegetatio*, 60, 113–18.

Green, G.M. and Sussman R.W., (1990) Deforestation history of the eastern rain forests of Madagascar from satellite images. *Science*, 248, 212–15.

Guichon, A. (1960) La superficie des formations forestieres a Madagascar. *Revue Forestière Française*, 6, 408–11.

Institut Geographique National (1968) *Madagascar et 'Mascareignes'*. Three sheets, 1:1 000 000 scale, Paris.

Institute National de Geodesie et Cattographie (1986). *Map of Madagascar*, 1:2 000 000, FTM.

IUCN/UNEP/WWF (1987) *Madagascar: An Environmental Profile* (ed. M.D. Jenkins), IUCN, Gland, Switzerland and Cambridge, UK, 374pp.

Luscombe, B.W. (1986) Spatial data handling in data-

* The World Bank Environment Department was reorganized in 1990 and these Unit names are therefore currently out-of-date.

poor environments. PhD dissertation, Department of Geography, Simon Fraser University, Vancouver.

May, R.M. (1992) The calculus of biodiversity. A presentation at the symposium entitled *Systematics and Conservation Evaluation* held at the Natural History Museum, London, 17–19 June 1992.

Miller, R.I. (1986) Predicting distribution patterns of vascular plants in the southern Appalachians of the southeastern United States. *Journal of Biogeography*, **13**, 293–311.

Miller, R.I. and White, P.S. (1986) Considerations for preserve design based on the distribution of rare plants in Great Smoky Mountains National Park. *Environmental Management*, **10**(1), 119–24.

Miller, R.I., Stuart, S.N. and Howell, K.N. (1989) A methodology for analyzing rare species distribution patterns utilizing GIS technology: The rare birds of Tanzania. *The Journal of Landscape Ecology*, **2**(3), 173–89.

Olson, S. (1988) Environments as shock absorbers, examples from Madagascar. *Environmental Review*, **12**(4), 161–80.

Petter, J.J., Albignae, R. and Rumpler, Y. (1977) *Mammiferes lemuriens* (Primates prosimiens). *Faune de Madagascar No. 44*, ORSTOM-CNRS, Paris.

Scott, J.M., Csuti, B., Jacobi, J.D. and Estes, J.E. (1987) Species richness: A geographic approach to protecting future biological diversity. *Bioscience*, **37**, 782–8.

Sullivan, T.A. (1992) Beyond hotspots: Assessing new approaches to setting priorities for the conservation of biodiversity. *Species*, **18**, 13–15.

Tattersall, I. (1982) *The Primates of Madagascar*, Columbia Univ. Press, New York.

United States Air Force (1967) *Operational Navigation Charts*. 1:1 000 000 scale, ONC-N6, P6, Q6, Revised May 1978, St. Louis Air Force Station.

Young, J.E. (1989) Madagascar teeters on the brink. *World Watch*, **2**(2), 10–12.

Chapter four

Modeling vertebrate distributions for Gap Analysis

Bart R. Butterfield, Blair Csuti and J. Michael Scott

4.1 Introduction

The number of species in imminent danger of extinction has grown throughout the world (Wilson and Peter, 1988). Though some local efforts have been successful, in general, traditional efforts to conserve individual species have demonstrated limited effectiveness. In particular, habitat protection is driven by the needs of a few species rather than the identification of habitats representative of all species. A new approach is needed for the conservation of nature. This new perspective must consider large regions, integrate the distribution of many species, and achieve rapid implementation. We have developed a prototype model in Idaho, USA, for a nationwide conservation evaluation project called Gap Analysis in response to these needs. This approach compares the distribution of existing vegetation and terrestrial vertebrates with areas managed for the long-term maintenance of biological diversity (Scott et al., 1987, 1989, 1993). Our goal is to identify the areas needed to complete a fully representative reserve network.

An integral component of Gap Analysis is the development of distribution maps for terrestrial vertebrates. Since detailed distributional records for most species were unavailable, we used a geographic information system (GIS) to create spatial models for the breeding distributions of 366 species of amphibians, reptiles, birds and mammals in Idaho. This chapter describes the development of those models and an example of their application in Gap Analysis.

Mapping the Diversity of Nature. Edited by Ronald I. Miller.
Published in 1994 by Chapman & Hall, London. ISBN 0 412 45510 2.

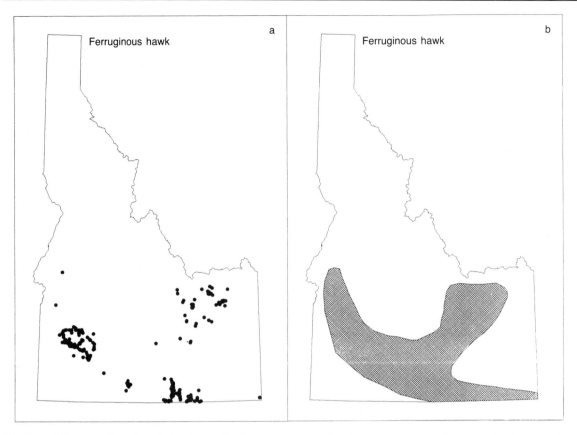

FIGURE 4.1 Examples of traditional distribution maps: (a) a dot distribution map of recorded sightings of the ferruginous hawk (*Buteo regalis*) in Idaho, (b) a range map drawn to enclose the dots in (a).

4.2 Methods

4.2.1 Background

Traditional species distribution maps are based upon reliable observations with specimen locations (Figure 4.1). Range maps are created by drawing boundaries around the location points of the dot-distribution maps. Range maps can also be made from the identification of existing geographic areas (e.g. grid squares or political regions) that contain specimen locations. However, these types of map are insufficient for use in Gap Analysis. The inadequacy of these map types for Gap Analysis

results because of coverage inconsistencies. For example, dot-distribution maps do not depict a species presence in areas that have not been searched or where a species is undetected. In addition, range maps encompass both appropriate and inappropriate habitat. Therefore, it is possible that some species may be absent from a large portion of their indicated range. These maps may result in erroneous conclusions if they are used to compare species distributions with patterns of land management. In order to meet the needs of Gap Analysis, distribution maps need to (1) depict the range limits of each species, (2) differentiate among areas where a species is likely to be present or absent within its range, (3) be reasonably accurate in the absence of extensive field inventory so that they can be used to address current conservation issues, (4) contain sufficient detail to allow comparison with land ownership and management patterns, and (5) not be overly complex (i.e. time and money resources need to be conserved).

For this prototype model we combined multiple layers of information that are correlated with terrestrial vertebrate species distributions. Two types of data are required for each layer used: (1) a map depicting the distribution of some environmental feature that is correlated to the distribution of vertebrates, and (2) a data file that associates presence or absence of species with the map elements. At least two map layers are used to model the distribution of each species: one layer depicts the known range limits and one or more layers depict the environmental features. The models operate in two steps. First, the known range of each species is selected. Then areas of appropriate habitat are identified within the selected range.

4.2.2 A simple model: County/vegetation

Our simplest vertebrate distribution model combines species specific occurrence by county with habitat preferences. A digital map of Idaho's 44 counties was created using maps from the US Geologic Survey (USGS) 1:100 000 scale map series. A data file that includes occurrences for each species in each county was obtained from the Idaho Conservation Data Center (formerly the Idaho Natural Heritage Program). County boundaries were determined to be the most tractable geographic unit for this pilot. However, counties are probably not the best geographic unit for depicting species ranges because: (1) county boundaries are artificial; (2) county areas vary widely, and (3) in Idaho county areas are too large to allow

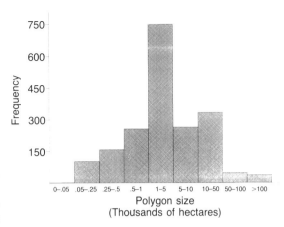

FIGURE 4.2 The frequency of polygon size distribution in the Gap Analysis vegetation map. The median polygon size in this map is 2537 ha.

detailed manipulation of the ranges of individual species.

The Gap Analysis vegetation map was compiled from existing large-scale vegetation maps and Landsat image interpretation. One-hundred and eighteen vegetation types are described on this map (Caicco, 1989). Each vegetation type is characterized by the dominant or codominant overstory plant species. These types were delineated on a 1:500 000 scale map of Idaho and digitized as the polygons of a GIS data layer. We created a data file that assigns presence or absence of each vertebrate species to each of the Gap Analysis vegetation types. We consulted several sources to determine each species' habitat preferences including: (1) the Idaho Conservation Data Center's Vertebrate Characterization Abstracts; (2) the scientific literature (Davis, 1939; Larrison and Johnson, 1981; Larrison *et al.*, 1967); and (3) the common species field guides (Burt and Grossenheider, 1964; Peterson and Peterson, 1990).

The Gap Analysis vegetation map demonstrated several limitations when we modeled vertebrate distributions. First, at the mapping scale of 1:500 000, the median polygon size is 2537 ha (Figure 4.2). Polygons this large are heterogeneous collections of vegetative communities in various seral stages. We assume that each vegetation polygon contains a variety of microhabitats required by vertebrates characteristic of the dominant vegetation type. A second limitation of the vegetation map is insufficient ecological information for some vegetation types (e.g. most notably agriculture and some sagebrush types). For example, we are unable to distinguish irrigated row crops from pasture. Despite these shortcomings, the Gap Analysis vegetation map remains the only complete map of existing vegetation for the entire state of Idaho and it is a prime component of the vertebrate distribution modeling in this project.

The vegetation and county map layers were overlaid in a GIS operation that created composite polygons that are identified by county and vegetation type. For each species, the GIS is instructed to extract the counties from the composite map layer when the species is coded as present. From this subset, the GIS then identifies all vegetation types when each species is coded as present. Species are modeled to be present in entire vegetation polygons even if the vegetation polygon extends across county lines into areas where the species is unrecorded. The resultant maps depict areas where each species is likely to be found in Idaho during the breeding season (Figure 4.3).

Paper maps were plotted for each species. A set of maps for each vertebrate class were examined by two acknowledged experts in Idaho. They marked errors on the maps and made suggestions for corrections. We adjusted the association tables appropriately and repeated the model operation. However, many of the evaluators' comments were not addressed by simple modifications in the association tables. Additional map layers that incorporated other important environmental factors that influence vertebrate distribution are required to refine the modeling effort.

4.2.3 Complex models

(a) County/vegetation/temperature

One group of species that elicit poor results with the county/vegetation model are reptiles. Their actual distributions are often more restricted than the distributions of the vegetation types and/or the counties with which they are associated. For example, the night snake (*Hypsiglena torquata*) is associated with sagebrush (*Artemisia tridentata*) communities (Nussbaum *et al.*, 1983). When county-of-occurrence and vegetation are extracted for this species and the distribution represented across county lines in the same vegetation polygon, the model for the night snake extends well beyond its documented range limits (Figure 4.4(a)). A factor other than vegetation appears to limit the distribution of the night snake and many other reptiles.

Reptiles are poikilothermic and they exhibit distributions strongly affected by temperature gradients, particularly in northern temperate regions such as Idaho. The western portion of the Snake River Plain is the warmest part of Idaho and many reptiles are found only in this area of the state. To the east of this area the average temperature decreases and sagebrush remains the dominant vegetation type. Therefore, temperature appears to be a likely variable for modeling reptile distributions in Idaho.

Weather-recording stations in Idaho are widely scattered and therefore sufficient weather data are not available for distribution modeling. In a previous study, Everson and Caprio (1974) prepared a detailed map that depicts the average date for the first bloom of lilacs (*Syringa vulgaris* L.) in Idaho using 10 years of state-wide data. This map is used as an indicator of temperature regimes in our models. Lilac development is strongly correlated to daily multiplication of solar radiation and mean daily temperature (Everson and Caprio, 1974).

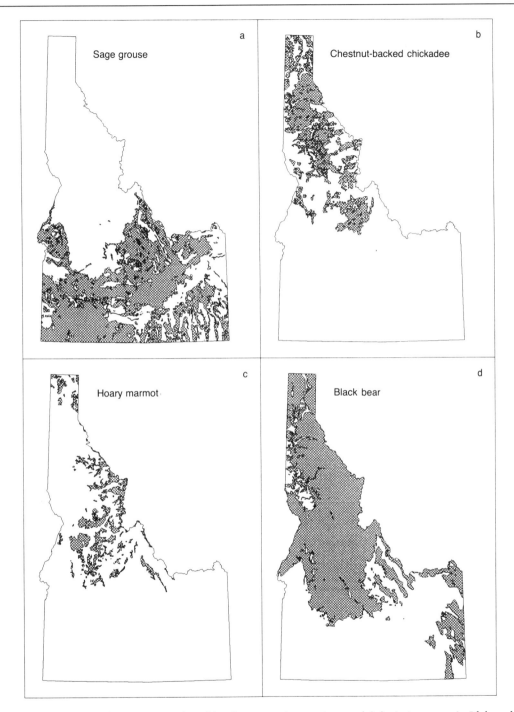

FIGURE 4.3 These are four maps produced by the county/vegetation model depicting areas in Idaho where each of these vertebrate species occurs during the breeding season: (a) sage grouse (*Centrocercus urophasianus*), (b) chestnut-backed chickadee (*Parus hudsonicus*), (c) hoary marmot (*Marmota caligata*), (d) black bear (*Ursus americanus*).

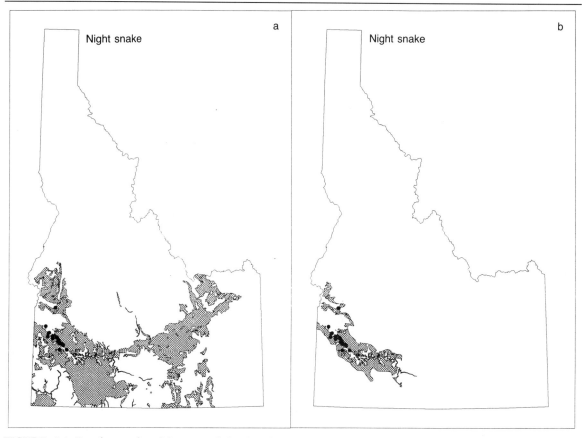

FIGURE 4.4 Results produced by two of the distribution models for the night snake (*Hypsiglena torquata*) in Idaho. The dots represent known locations of this species. The model that produced (a) uses only county-of-occurrence and vegetation association data. The model that produced (b) uses an additional temperature layer that is mapped as the date of first bloom of lilacs.

To test the importance of temperature, we digitized the lilac-blooming map and combined it with the county and vegetation maps to create a GIS map layer with composite polygons. We then used both the dot-distribution maps (Nussbaum *et al.*, 1983) and the pertinent comments from our initial distribution maps to correlate the distribution of each reptile species with the date of first bloom of lilacs. The county/vegetation/temperature model repeats the steps of the county/vegetation model and further restricts the species distribution models with the associated lilac phenology map. The resulting distribution maps more closely corres-

pond to the known ranges of reptiles in Idaho (Figure 4.4(b)).

(b) County/vegetation/riparian–wetlands

Vertebrate species associated with riparian or wetland habitats are another difficult group to model using county and vegetation variables. Most riparian habitat in Idaho is restricted to narrow corridors along streams. Similar water-dependent vegetation around lakes and small wetland areas also attract riparian or wetland-associated species. We were able to depict only the largest of these important habitats in the Gap Analysis vegetation map. A model of the

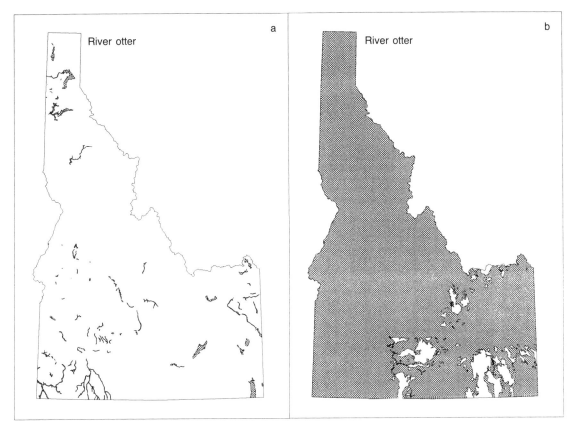

FIGURE 4.5 The mapped results of two distribution models for the river otter (*Lutra canadensis*) in Idaho. The distribution in (a) underestimates the actual distribution pattern of the species because the species uses riparian habitat that does not appear in the Gap Analysis vegetation map at the 1:500 000 scale. The distribution in (b) depicts expansive uplands when riparian habitats are assumed in otherwise appropriate vegetation types.

distribution of riparian and wetland species on our map of vegetation types depicts these species in only a few areas (Figure 4.5(a)). On the other hand, if we assume that riparian and wetland habitats occur as microhabitats in vegetation polygons, we map those same species across broad expanses of uplands (Figure 4.5(b)).

We modeled riparian habitat by creating 100–400 m buffers around all streams and lakes in the USGS 1:100 000 scale digital hydrography data. These data include all perennial streams and lakes in Idaho and the data density make it unmanageable with the computer hardware and software available to

us. Therefore, we reduced the hydrography data density by eliminating most of the Strahler (1957, 1964) first-order streams. This effectively increased the mean distance between headwater streams from approximately 300 m to about 1 km. We then adjusted the riparian buffers to reflect the size class of their associated stream or lake. We used all the wetland symbology on the USGS 1:100 000 scale map series for our wetlands map. We also digitized additional areas listed as important wetlands in Idaho by the Idaho Department of Fish and Game, US Environmental Protection Agency, and US Fish and Wildlife Service.

The riparian and wetland maps were then

combined with the pre-existent map layers to create a composite GIS map layer of counties, vegetation types, dates of first blooms for lilacs, riparian habitats, and wetlands. This model repeats the county/vegetation model procedures and it further restricts distributions to the correct class of riparian or wetland habitats (Figure 4.6(a)).

Each of the GIS map layers include inadequacies that are attributable to both elements of the data-processing methods and to the diversity of the natural conditions. Since we needed to reduce the hydrography data before buffering and modeling, we did not represent all the riparian habitat in Idaho. Similarly, we were unable to delineate all of the small patches of wetland habitat at the 1:100 000 mapping scale. The vertebrate species with distributions most accurately modeled by the county/vegetation/riparian–wetland model are those species that use riparian or wetland habitat that can be mapped at this scale.

Northern orioles (*Icterus galbula*) and song sparrows (*Melospiza melodia*) are two species that are both riparian habitat specialists in Idaho. The distribution of northern orioles is modeled with the county/vegetation/riparian–wetland model because this species is generally restricted to larger streams and lake shores, a habitat included in the model (Figure 4.6(b)). Song sparrows, however, are found in even the smallest riparian strips along headwaters and intermittent streams and therefore the most realistic model for song sparrows is the county/vegetation model (Figure 4.6(c)). As a general rule, we apply the county/vegetation/riparian–wetland model to species associated with larger riparian and wetland areas, but we use the more general county/vegetation model for more ubiquitous species.

The creation of a single map permits us to consider the changing habitat variables that are important to individual species or groups of species. This is achieved by the manipulation of a composite map produced from the consolidation of the polygons in each of the constituent base layers. For example, some vertebrate species in Idaho use different habitats in different parts of the state. The northern saw-whet owl (*Aegolius acadicus*) generally uses upland forests in the northern and central portions of the state, but it is restricted to riparian habitat in the southern shrub-steppe regions. To produce a distribution model for this species we apply the general county/vegetation model where the owl occurs in upland forests and we use the county/vegetation/riparian–wetland model where the owl occurs in shrub-steppe regions (Figure 4.6(d)).

(c) County/vegetation/potential natural vegetation

We were unable to map certain floristic elements important to some vertebrate species although we identified 118 detailed vegetation types in Idaho. For example, sagebrush communities that occur across much of southern Idaho have an understory of bunchgrasses. However, in the increased elevation regime of southeastern Idaho, these communities support an understory of forbs and fruit-bearing shrubs used by the sharp-tailed grouse (*Tympanuchus phasianellus*). This understory flora is captured in a map of potential natural vegetation produced by the US Soil Conservation Service (SCS). This SCS map is combined with the other base map layers to produce a single, complex map layer that constitutes the county/vegetation/potential natural vegetation model. It executes by first repeating the county/vegetation model and then identifying potential natural vegetation types that are associated with particular species (Figure 4.7(a)).

(d) County/vegetation/faunal regions

With the previous model types we are unable to adequately model the distribution of a handful of fossorial rodents. We know that soil characteristics influence the distribution of certain

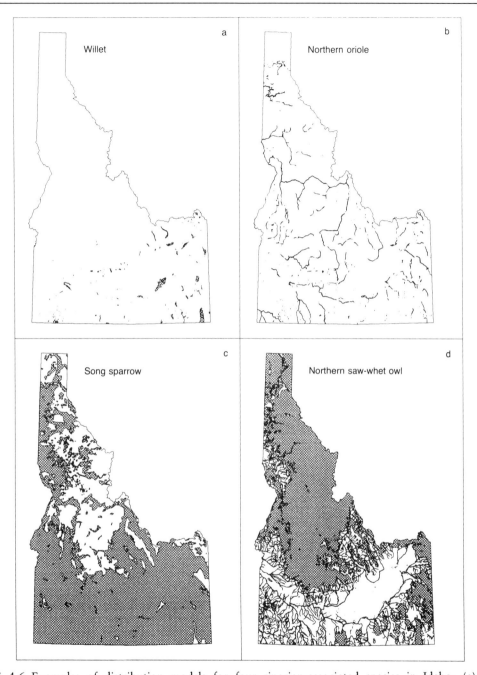

FIGURE 4.6 Examples of distribution models for four riparian-associated species in Idaho: (a) the willet (*Catoptrophorus semipalmatus*) and (b) northern oriole (*Icterus galbula*) are depicted using the county/vegetation/riparian–wetland model. (c) The song sparrow (*Melospiza melodia*) uses riparian habitats at too fine a scale to depict its distribution for the entire state. The song sparrow distribution is successfully depicted using the county/vegetation model. (d) The northern saw-whet owl (*Aegolius acadicus*) uses upland forests where available but it is restricted to riparian habitats in shrub steppe regions. The northern saw-whet owl distribution is best depicted using a combination of models.

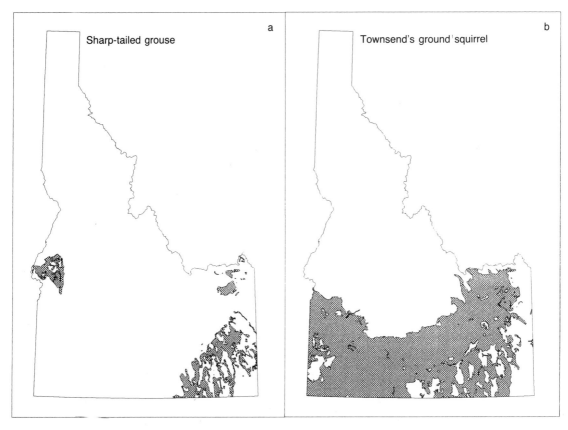

FIGURE 4.7 Two examples of species distributions that result from (a) the county/vegetation/potential natural vegetation model and (b) the county/vegetation/faunal region model. The species represented are (a) sharp-tailed grouse (*Tympanuchus phasianellus*) and (b) Townsend's ground squirrel (*Spermophilus townsendii*).

burrowing rodents, but a state-wide soils map was unavailable. (Since our initial model development the US Soil Conservation Service has completed a state-wide soils map for Idaho at a scale of 1:250 000.) Some major rivers appear to define range boundaries for these rodents (Larrison and Johnson, 1981) and Davis (1939) used these ranges to define faunal regions in Idaho. It is still not clear whether the rivers are acting as barriers to faunal dispersal or whether the river locations are correlated with some other factor that influences the faunal ranges. We used the USGS 1:100 000 scale digital hydrography data to produce faunal regions from river boundaries in areas with endemic

rodent species. This was similar to the approach used by Davis (1939). This model, designated as the county/vegetation/faunal region model, then directs the application of the county/vegetation model to specific regions for designated vertebrate species (Figure 4.7(b)).

(e) Rare and sensitive species

Rare and sensitive species generally occur only in a small amount of the appropriate habitat within their range. They may be found at a few separate sites or be scattered sparsely across their range. Conservation concerns require an accurate depiction of the distributions of rare

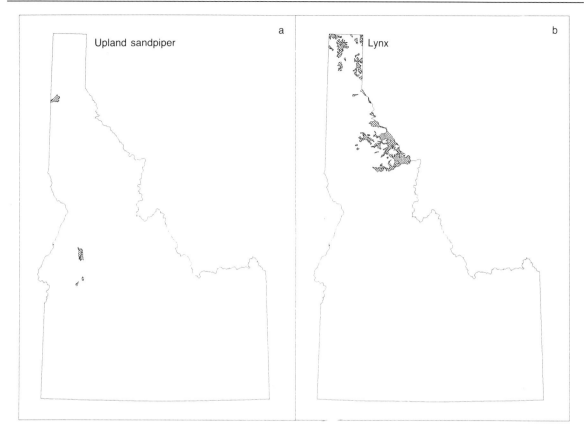

FIGURE 4.8 These two maps have been produced for rare and sensitive species in Idaho. The points of known occurrence are used in (a) to identify the occupied habitat polygons for the upland sandpiper (*Bartramia longicauda*). An interactive modification of the county/vegetation model is used in (b) to incorporate the best-known professional judgement for the extent of lynx (*Felis lynx*) in Idaho.

and sensitive species in Gap Analysis. Ironic-ally, the rare and sensitive species are often better inventoried than more common species. The Idaho Conservation Data Center collects and catalogs locality information for many of these rare and sensitive species.

We imported the Idaho Conservation Data Center's rare and sensitive species database into the Gap Analysis GIS as a set of points. During the process we omitted observations recorded previous to 1950 and observations not recorded within a species breeding season. Each rare and sensitive species is assigned to one of the previously described model types, according to

its general habitat preferences (e.g. riparian species to the county/vegetation/riparian–wet-lands model). The appropriate model is run for each species and polygons are identified only if they contain location points for the species. The distribution maps consist of polygons of appro-priate habitat that contain species observations (Figure 4.8(a)).

Rare, wide-ranging carnivores, such as wolver-ine (*Gulo gulo*), lynx (*Felis lynx*), fisher (*Martes pennanti*), wolf (*Canis lupus*) and grizzly bear (*Ursus arctos*), have home ranges that are often larger than a single vegetation polygon. They also occupy remote wildlands where their presence is unpredictable. Consequently, the use of point observations for these species produces distribution models smaller than their known current ranges. The county/vegetation model can be used for these species, but it depicts these species to be present in areas outside of their currently known distribution where they have been extirpated. We used the GIS to interactively edit the distributions produced by the county/vegetation model to agree with our best professional judgement about the distribution attributes of each of these species (Figure 4.8(b)).

4.3 Application to Gap Analysis: An example

Gap Analysis identifies vertebrate species that are not adequately represented within areas managed for the long-term maintenance of biological diversity. Further, it identifies and maps areas with high numbers of these species. In this example, the two map layers required for analysis are vertebrate species distributions and special management areas. The vertebrate distributions and their creation are discussed above. The special management areas map was compiled from the Idaho Conservation Data Center's managed area database.

For Gap Analysis, a special management area must have a management plan in operation which maintains the area in its natural state for the foreseeable future. Alternative land uses are acceptable as long as the primary goal is met for the long-term maintenance of biological diversity. Examples of special management areas include: national parks, wilderness areas, wildlife refuges, US Forest Service research natural areas, US Bureau of Land Management areas of critical environmental concern, and some state parks and privately owned preserves. Public lands in general are not considered special management areas for Gap Analysis because resource use and production

activities often reduce local and regional biological diversity.

Analysis begins with the overlay of the special management areas with the vertebrate distribution models. The occurrence of each species within the special management areas is then calculated in relation to predetermined criteria. For example, in the Idaho prototype, a species was considered adequately protected if its distribution model included at least 10 000 contiguous hectares inside special management areas. Species failing to meet this test are the 'gaps' not adequately represented in the existing network of special management areas (Burley, 1988) (Table 4.1). The distribution models of these species were combined in a GIS additive overlay procedure. Areas rich in 'gap' species have high numbers of species not adequately represented in existing special management areas (Figure 4.9). These areas require further intensive study to determine their suitability for land management practices that would enhance long-term maintenance of biological diversity.

4.4 The final implications

4.4.1 Conclusions

Gap Analysis in Idaho identifies areas rich in species that are not currently represented in areas managed for the long-term maintenance of biological diversity. These areas could be important contributions to a reserve network that would be fully representative of all species and ecosystems. Input into this Gap Analysis project included breeding distribution data for 366 species of amphibians, reptiles, birds and mammals. These species distributions are modeled, based upon their known range limits and their association with mappable features of the environment. Eight map layers were used in a GIS to produce these models:

1 County of occurrence
2 Vegetation type

TABLE 4.1 List of vertebrate species that breed in Idaho that do not have occurrences of 10 000 or more continuous hectares inside a special management area[a]

Wood frog	*Rana sylvatica*	American avocet	*Recurvirostra americana*
Painted turtle	*Chrysemys picta*	Willet	*Catoptrophorus semipalmatus*
Northern alligator lizard	*Elgaria coerulea*	Upland sandpiper	*Bartramia longicauda*
Mojave black-collared lizard	*Crotaphytus bicinctores*	Wilson's phalarope	*Phalaropus tricolor*
Desert horned lizard	*Phrynosoma platyrhinos*	Franklin's gull	*Larus pipixcan*
Side-blotched lizard	*Uta stansburiana*	Ring-billed gull	*Larus delawarensis*
Western whiptail	*Cnemidophorus tigris*	California gull	*Larus californicus*
Striped whipsnake	*Masticophis taeniatus*	Caspian tern	*Sterna caspia*
Common loon	*Gavia immer*	Common tern	*Sterna hirundo*
Pied-billed grebe	*Podilymbus podiceps*	Yellow-billed cuckoo	*Coccyzus americanus*
Horned grebe	*Podiceps auritus*	Black swift	*Cypseloides niger*
Red-necked grebe	*Podiceps grisegena*	Gray flycatcher	*Empidonax wrightii*
Eared grebe	*Podiceps nigricollis*	Ash-throated flycatcher	*Myiarchus cinerascens*
Western grebe	*Aechmophorus occidentalis*	Scrub jay	*Aphelocoma coerulescens*
Clark's grebe	*Aechmophorus clarkii*	Pinyon jay	*Gymnorhinus cyanocephalus*
American white pelican	*Pelecanus erythrorhynchos*	Boreal chickadee	*Parus hudsonicus*
Double-crested cormorant	*Phalacrocorax auritus*	Plain titmouse	*Parus inornatus*
American bittern	*Botaurus lentiginosus*	Marsh wren	*Cistothorus palustris*
Great egret	*Casmerodius albus*	American dipper	*Cinclus mexicanus*
Snowy egret	*Egretta thula*	Blue-gray gnatcatcher	*Polioptila caerulea*
Cattle egret	*Nubulcus ibis*	American pipit	*Anthus spinoletta*
Black-crowned night heron	*Nycticorax*	Red-eyed vireo	*Vireo olivaceus*
White-faced ibis	*Plegadis chihi*	Virginia's warbler	*Vermivora virginiae*
Trumpeter swan	*Cygnus buccinator*	Black-throated gray warbler	*Dendroica nigrescens*
Northern pintail	*Anas acuta*	American redstart	*Setophaga ruticilla*
Blue-winged teal	*Anas discors*	Northern waterthrush	*Seiurus noveboracensis*
Northern shoveler	*Anas clypeata*	Blue grosbeak	*Guiraca caerulea*
Gadwall	*Anas strepera*	Black-throated sparrow	*Amphispiza bilineata*
American wigeon	*Anas americana*	Yellow-headed blackbird	*Xanthocephalus xanthocephalus*
Canvasback	*Aythya valisineria*	Scott's oriole	*Icterus parisorum*
Redhead	*Aythya americana*	Rosy finch	*Leucosticte arctoa*
Ring-necked duck	*Aythya collaris*	Lesser goldfinch	*Carduelis psaltria*
Lesser scaup	*Aythya affinis*	Coast mole	*Scapanus orarius*
Common goldeneye	*Bucephala clangula*	Fringed myotis	*Myotis thysanodes*
Barrow's goldeneye	*Bucephala islandica*	California myotis	*Myotis californicus*
Bufflehead	*Bucephala albeola*	Cliff chipmunk	*Tamias dorsalis*
Hooded merganser	*Lophodytes cucullatus*	Uinta chipmunk	*Tamias umbrinus*
Common merganser	*Mergus merganser*	White-tailed antelope squirrel	*Ammospermophilus leucurus*
Ruddy duck	*Oxyura jamaicensis*	Idaho ground squirrel	*Spermophilus brunneus*
Sharp-tailed grouse	*Tympanuchus phasianellus*	Rock squirrel	*Spermophilus variegatus*
Northern bobwhite	*Colinus virginianus*	Townsend's pocket gopher	*Thomomys townsendii*
Gambel's quail	*Callipepla gambelii*	Little pocket mouse	*Perognathus longimembris*
Virginia rail	*Rallus limicola*	Dark kangaroo mouse	*Microdipodops megacephalus*
Sora	*Porzana carolina*	Chisel-toothed kangaroo rat	*Dipodomys microps*
American coot	*Fulica americana*	Pinon mouse	*Peromyscus truei*
Sandhill crane	*Grus canadensis*	Northern bog lemming	*Synaptomys borealis*
Whooping crane	*Grus americana*	Lynx	*Felis lynx*
Black-necked stilt	*Himantopus mexicanus*	Caribou	*Rangifer tarandus*

[a] These vertebrates are listed in order of a taxonomic sequence system established by the Heritage Program of The Nature Conservancy.

65

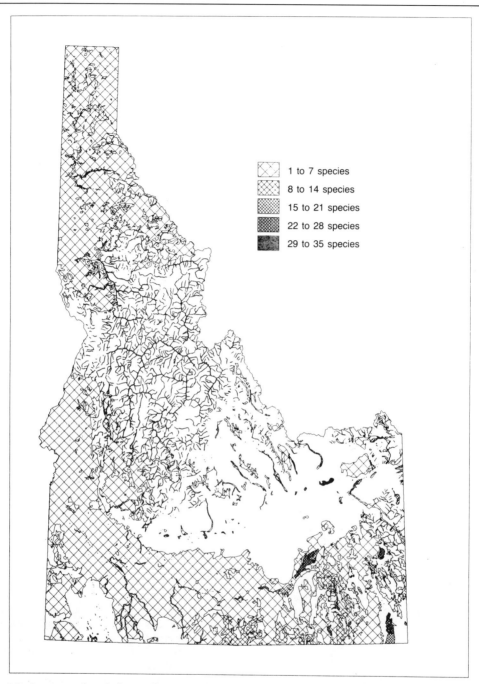

⊠	1 to 7 species
⊠	8 to 14 species
⊠	15 to 21 species
⊠	22 to 28 species
■	29 to 35 species

FIGURE 4.9 An example of the application of vertebrate distribution models in Gap Analysis. Terrestrial vertebrate breeding distributions (366 species) are overlaid onto a map of the special management areas. Any species without at least one occurrence of 10 000 contiguous hectares in a special management area is tagged as inadequately protected. Based upon these criteria, the distributions of these 'gap' species are additively overlaid to produce this richness map for the species inadequately protected in Idaho. The denser shade patterns represent higher numbers of species; lighter shade patterns represent lower number of species.

3 Date of first bloom of lilacs (as an indicator of temperature)
4 Stream and lake riparian habitats
5 Wetlands
6 Potential natural vegetation
7 Faunal regions
8 Location points of rare and sensitive species.

4.4.2 Assumptions and limitations

The following fundamental assumptions are necessary in the compilation and application of the Gap Analysis vertebrate distribution models:

Vertebrate species respond to environmental features that can be represented on a map. These features constitute the map layers used in this study. The relationship of species to the environment is a major focus of modern ecological research. Although factors such as interspecific competition may play some role in shaping species distributions, the importance of the physical environment cannot be underestimated. The data underpinning the relationship between species and the environment can be obtained in varying degrees from scientific literature, popular accounts, and from unpublished communications between reliable observers.

Observations of species outside their normal range or habitat are considered vagrant or migratory and these are not used as input for the distribution models. Many species, especially birds and large mammals, are highly mobile and wander widely. Tropical birds are occasionally seen in mid-latitudes, marine birds sometimes occur far inland, and ungulates are sometimes seen wandering through large towns in Idaho. We consider these observations to be atypical and we do not use them in the distribution models.

Polygons in the map layers contain microhabitats necessary for the occurrence of many of the species. Most species of vertebrates are

adapted to some special microhabitat feature. For example, for nesting and feeding, many birds use snags, reptiles often use rocky areas or sandy spots, and rodents often require particular soil characteristics. These are features at a fine grain of ecological resolution and they cannot be mapped at our scale of analysis. We assume the presence of these microhabitats in otherwise appropriate areas even when their locations are not explicitly documented.

The distribution of individuals within polygons produced by the models is continuous. For production and display of the distribution models this assumption is not necessary. In fact, the actual distribution of individuals in nature is not continuous. Nevertheless, Gap Analysis assumes that species occur at all points in their modeled distributions. Otherwise, the modeled presence of a species in a special management area could not be definitively ascertained.

The data used in this prototype Gap Analysis project also have certain limitations:

1 The minimum mapping unit for the Idaho Gap Analysis vegetation map is approximately 200 ha and the median polygon size is 2537 ha (Figure 4.2). This makes the data appropriate for large area analyses, but they should not be used for stand level evaluations.
2 The distribution maps represent the results of a modeling effort and they are not documented observations. The models certainly contain some error though they are based upon sound, scientifically established relationships with environmental features. We are currently analyzing the data to quantify the error in the distribution models and preliminary results indicate that, when compared to species lists compiled on national wildlife refuges, omission and commission errors are less than 30% (Scott *et al.*, 1993).

3 The distribution maps depict species presence. Population density is not reflected in the models.

4.5 Summary

The loss of native species and habitat has reached crisis proportions worldwide. Immediate conservation action is required, but is hampered by a single species focus, a lack of knowledge about species distribution, and local rather than regional perspectives. In Idaho, USA we have developed spatial models of the breeding distributions of 366 species of amphibians, reptiles, birds and mammals. The models are based upon the known range limits of each species and their association with mappable environmental features. The models are applicable to regional planning and analysis, but they are not applicable at the local or stand level. Similar state-wide models are being developed across the United States as part of a nationwide conservation evaluation program called Gap Analysis. The state-wide models will be joined to provide detailed distribution models of vertebrates, and will be useful for regional or national conservation planning and analysis.

References

Burley, F.W. (1988) Monitoring biological diversity for setting priorities in conservation, in *Biodiversity* (ed. E.O. Wilson and F.M. Peter), National Academy Press, Washington, D.C., pp. 227–30.

Burt, W.H. and Grossenheider, R.P. (1964) *A Field Guide to the Mammals*, Houghton Mifflin Co., Boston.

Caicco, S. (1989) *Manual to Accompany the Map of Existing Vegetation of Idaho*. Unpublished manuscript, Idaho Cooperative Fish and Wildlife Research Unit, University of Idaho, Moscow.

Davis, W.B. (1939) *The Recent Mammals of Idaho*, Caxton Printers, Caldwell, Idaho.

Everson, D.O. and Caprio, J.M. (1974) *Phenological Map of Average Date when Lilacs Start Bloom in Idaho*, Miscellaneous Publication No. 18, Idaho Agricultural Experiment Station, University of Idaho, Moscow.

Larrison, E.J., Tucker, J.L. and Jollie, M.T. (1967) *Guide to Idaho Birds*, Idaho Academy of Science, University of Idaho, Moscow.

Larrison, E.J. and Johnson, D.R. (1981) *Mammals of Idaho*, The University of Idaho Press, Moscow.

Nussbaum, R.A., Brodie, E.D. and Storm, R.M. (1983) *Amphibians and Reptiles of the Pacific Northwest*, University Press of Idaho, Moscow.

Peterson, R.T. and Peterson, V.M. (1990) *A Field Guide to Western Birds*, Houghton Mifflin Co., Boston.

Scott, J.M., Csuti, B., Jacobi, J.J. and Estes, J.E. (1987) Species richness: a geographic approach to protecting future biological diversity. *BioScience*, 37, 782–8.

Scott, J.M., Csuti, B., Estes, J.E. and Anderson, H. (1989) Status assessment of biodiversity protection. *Conservation Biology*, 3, 85–7.

Scott, J.M., Davis, F., Csuti, B. *et al.* (1993) Gap analysis: a geographic approach to protection of biological diversity. *Wildlife Monographs*, No. 123.

Strahler, A.N. (1957) Quantitative analysis of watershed geomorphology. *American Geophysical Union Transactions*, 38, 913–20.

Strahler, A.N. (1964) Quantitative geomorphology of drainage basins and channel networks, in *Handbook of Applied Hydrology, Sect. 4-II* (ed. V.T. Chow), McGraw-Hill, New York, pp. 4-39–4-76.

Wilson, E.O. and Peter, F.M. (eds) (1988) *Biodiversity*, National Academy Press, Washington, D.C.

Part Three

A Conceptual Context for Biodiversity Mapping

Most of the data that document the locations of species around the world are geographically coarse while GIS technology permits the generation of detailed and complex representations of species distributions. For example, different representations of species distributions can be displayed when these data are differentiated along axes that represent the dimensions within which these data reside. Chapter 5 classifies these different dimensions in connection with different data extents, storage schemes, themes and time frames. This is an excellent conceptual framework for understanding the representation of species data in regard to the many dimensions within which these data must be considered. This arrangement permits a comparison of multiple representations of disparate species and habitat data and it can produce more accurate species distributions from the convergence of the evidence. The implementation of these concepts are presented with an example using the distribution of a lizard, the orange-throated whiptail (*Cnemidophorus hyperythrus*), that is native to southern California. Data sources include generalized range outlines, museum records, field observations, climate data, vegetation maps and satellite imagery. The final analysis first maps overall lizard range limits, then the suitable habitats within the range, and finally the habitats within a local area. The approach introduced in Chapter 5 provides a clearer view of the data and permits us to make new inferences about the relationships between species and habitats.

Chapter five

Hierarchical representations of species distributions using maps, images and sighting data

Allan D. Hollander, Frank W. Davis and David M. Stoms

5.1 Introduction

Recent advances in automated systems that handle spatial data have revolutionized the way we represent distributions of plant and animal species. The traditional dot maps, grids or line drawings used until very recently to show species ranges are being replaced by applications of models to maps of habitat variables (e.g. Walker, 1990; Scott *et al.*, 1993). Models of ranges of populations or metapopulations now take explicit account of spatial effects, e.g. patch size, distance between suitable habitat patches, and the qualities of contiguous habitats (e.g. Hanski and Gilpin, 1991). Models of the behavior and movements of individual organisms now are carried out on spatial grids which accommodate effects such as territoriality and foraging (e.g. Folse *et al.*, 1989). The range of environmental data to support these modeling efforts has also grown exponentially, due to technological advances (i.e. improvements in digitizing, satellite and aircraft remote

Mapping the Diversity of Nature. Edited by Ronald I. Miller.
Published in 1994 by Chapman & Hall, London. ISBN 0 412 45510 2.

sensing, telemetry and Global Positioning Systems).

These new opportunities have created a host of new scientific and technical challenges. These challenges are currently defined by the limitations of extant ecological theory, data sources, and data handling techniques. Current ecological and biogeographical theories explain the distribution patterns of species and the processes responsible for them but they do not address the spatial patterns and processes that emerge at the different levels of aggregation from individuals to populations to subspecies and species (Levin, 1992). We lack even rudimentary data on distribution and habitat requirements for most species. Data are available for a handful of taxa of special interest, and these data are very localized in space and time. Few existing database systems are optimized for the maintenance and integration of biological data from diverse sources and map scales. Therefore these data are currently poorly coupled to analytical tools for modeling, visualization and uncertainty analysis.

This chapter is primarily concerned with design of biological databases for predicting species distributions. The fundamental premise of the approach presented here is that biodiversity data are naturally organized by theme, spatial extent, spatial grain (resolution), temporal extent and temporal grain (Figure 5.2). We focus here on the concept of spatial hierarchy, and we illustrate the use of a multitiered database to represent the distribution of the orange-throated whiptail (*Cnemidophorus hyperythrus*), a teiid lizard found in southern California and Baja California. We demonstrate the application of spatial scale with a distribution model constructed with the use of field data, multiresolution satellite imagery, generalized habitat maps, digital climatic data and range data. This approach presents suites of distribution maps at three different extents: the biogeographic, regional and local (Wiens, 1989; Davis *et al.*, 1990).

Together these three extents create an informative picture of the geography of the species in southern California, and the ecological factors associated with its known distribution. Data assembled from different sources also reveal the limitations and potential biases of existing information on the orange-throated whiptail. The sections that follow present a brief background, the biogeographic, regional and local perspectives, and a discussion of suggested refinements to our database model.

5.2 Conceptual background

5.2.1 Species distribution data

Today the basic challenge in distribution mapping is to be able to generalize sparse species data originally collected at fine spatial and temporal grains. There are a number of approaches to this age-old problem of geographic interpolation (e.g. Rapoport, 1982). To make an accurate analysis of species occurrence data, it is important to distinguish the differences between data extent (i.e. the total area analyzed), data grain and sampling intensity (fraction of extent sampled at a given grain) (Turner *et al.*, 1989). For example, field data are the most concrete data for reconstructing the range of a species (biogeographic extent). However, these data are often patchy, of indeterminate grain, and only cover a small fraction of the total extent.

In our view, a simple plot of locality data on maps or grids is the most objective but least informative representation of distribution. A much more useful representation is one where locality data with a consistent grain are displayed upon a map with species' occurrences associated with ecological requirements. This type of map would be derived from distributional maps of environmental variables.

At present the greatest difficulty that hampers such a probabilistic representation is

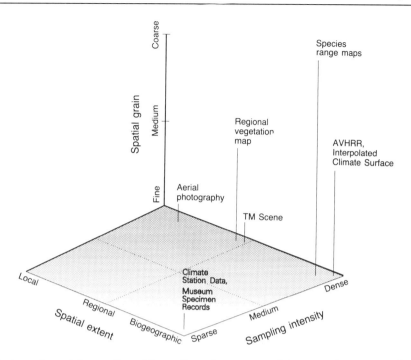

FIGURE 5.1 Scale and spatial sampling. Three dimensions are used here to characterize the scale of spatial observations, these being spatial grain, spatial extent and sampling intensity. A variety of biodiversity data types are plotted with respect to these axes. The figure illustrates how these data span much of the space defined by the axes. Comparison of data sets is facilitated by relative closeness in this space.

the lack of scientific understanding of the factors that influence the distribution patterns of species, populations and individuals. Most significantly, these factors vary in importance at different spatial and temporal grains. For plants, Woodward (1987) suggests that the temporal dynamics of species are dependent upon the relation between scale of environmental variation (e.g. seconds, days, years, centuries) and the time scale of organismal response. Specifically, a species will be influenced by those changes in the environment that occur over time intervals equal to or greater than the organism's response time. Likewise, a similar principle may apply to spatial variation among mobile organisms. Consequently, a taxon responds to spatial variation at or above the spatial grain size that it senses. This represents the home ranges of individual animals or the colonization ranges of

populations. In the absence of a strong predictive model based on these factors, a rule-based model based on more readily observed surrogates provides an interim representation that can be refined as data and understanding improve.

5.2.2 Biodiversity data sets and the database hypercube

Biodiversity data sets vary greatly, both spatially and temporally, in their properties of data extent, data grain and sampling intensity. Data sets differing in these properties will be difficult to compare. In general, a comparison is accomplished by extrapolation or interpolation of one of the data sets. Figure 5.1 presents a typical set of biodiversity data sets plotted along the three axes of spatial extent, spatial grain and sampling intensity. The figure points out the disparity of the placement of these data sets along these different axes. Choropleth maps are difficult to situate in this diagram in terms of 'sampling intensity' since they are a derived data set; the convention taken here is that the intensity is dense since polygons in the map exhaustively cover the spatial extent. The figure shows how transformation of datasets leads to changes in position in this space. For instance, climate station data of fine grain and sparse sampling are interpolated to form a map with exhaustive coverage and coarse grain.

An alternative, less abstract presentation of biodiversity data is possible in terms of four axes representing different data types, spatial extents, tiling schemes and temporal extents (i.e. time). Figure 5.2 (after Berry, 1964) graphically represents these axes in terms of a hypercube. Different data types can include maps of cultural features (e.g. land use and ownership patterns), different habitat elements (e.g. vegetation patterns), direct information on species distributions (e.g. sighting records) and remotely sensed data (e.g. aerial photographs, satellite images). These data can vary in spatial extent, ranging from a map of the entire United States to a local zoning map. The data can be compiled using different tiling schemes, such as political units (e.g. the counties of a state) or by grid units (e.g. the US Geological Survey 7.5′ quadrangle maps). These same data will vary temporally and reflect changing environmental patterns (e.g. landscape succession dynamics). The rate of change over time will depend upon the type of data being considered; political boundaries are more stable than land-use categories. Sampling intensity is not made explicit in this representation. Also, since the data grain is usually correlated with the data extent, these two factors are not distinctly categorized.

Most analyses will be sensitive to the position that the data occupy in the hypercube. Many difficulties encountered during the analysis of biodiversity data are explicated using the

hypercube presentation. For example, difficulties using data collected at different scales are more clearly differentiated in this depiction. Other difficulties include comparison of data collected at different time periods, using different sampling methodologies, and using different classification systems. All these complications are more clearly understood from the hypercube presentation. The hypercube therefore is an attempt to represent the many elements of variation in biodiversity data that need to be addressed in any graphical representation.

The following analysis of the whiptail distribution is structured along three different extents. We label these the biogeographic extent, the regional extent and the local extent. The biogeographic extent ranges in length from hundreds to thousands of kilometers, roughly corresponding to the size of the range of a typical terrestrial vertebrate. The regional extent ranges in length from tens to hundreds of kilometers. The local extent has a length of several kilometers. A model of home range and individual movements for many animals would necessitate a grain size of meters, but our data preclude such an analysis. Since we employ a simple statistical analysis rather than attempt to model dynamic processes, our representations across scales are accomplished either by spatial constraint or extrapolation (Turner *et al.*, 1989). Therefore the model developed at a coarse extent sets the spatial limits over which the model at a finer extent applies. Or a model developed over a small region is assumed to hold over a larger extent, and the larger region is similarly mapped.

The identification of the 'best' correlative model is difficult. Usually data from species observations are not systematically collected, and these data are seldom geographically well-referenced. Therefore, a rigorous analysis of a species distribution is usually not possible, since hypotheses cannot be accurately tested using these data. But a qualitative notion of the distribution of a species can be constructed

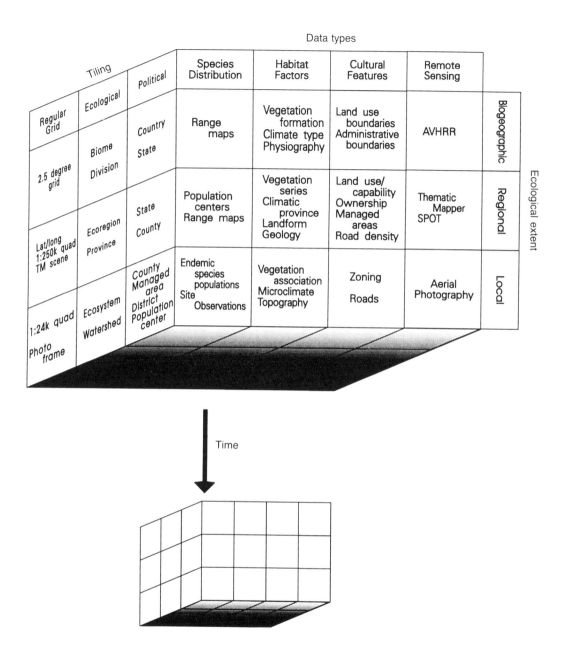

FIGURE 5.2 The biodiversity database hypercube. Data pertaining to the assessment of biodiversity are explicitly categorized along four axes of data types, ecological extent, tiling scheme and time of collection. Implicit in this representation are the frequency of data collection (spatial and temporal) and the data grain, which is generally correlated with extent. The following terms used in this hypercube are defined in the following references: vegetation formation, series and association (Parker and Matyas, 1981; Shimwell, 1971), ecoregion (Bailey, 1989), province (Daubenmire, 1978). (After Berry, 1964.)

without the usual accompanying statistical support through exploratory data analysis and convergence of evidence. This analytic approach uses GIS technology to visualize many exploratory models simultaneously.

5.3 Analysis and visualization of the orange-throated whiptail distribution

5.3.1 Whiptail natural history

We chose the orange-throated whiptail (*Cnemidophorus hyperythrus*) for this exercise in distribution mapping due to data availability and because it is a species of concern in an area threatened with much urban development. This is a teiid lizard that ranges from coastal southern California west of the Peninsular Ranges to the southern end of Baja California. In this analysis, we are only concerned with the northern portion of the range that falls in California, as illustrated in Figure 5.3. Its habitat is sparsely vegetated slopes or washes with open, heterogeneous brush and friable soils for burrowing. The whiptail focuses on termites but prey species can also include a variety of arthropods. Bostic (1965) speculated that the distribution of *C. hyperythrus* might be limited to a coincidence with one particular species of termite (Stebbins, 1985; Bostic, 1965).

The whiptail exhibits a range of diverse natural history characteristics. The mean home range is small, about 0.1 acres. The species is seasonally inactive. For example, in San Diego County, adult whiptails entered hibernation from late July through September, while immatures entered hibernation by December. Adults and immatures emerged around March or April of the following year. During the summer, individuals are most active in the mid-morning and the mid-afternoon. At mid-day during this same season, most individuals retreat to the shade of bushes or burrows (Bostic, 1966). Knowledge

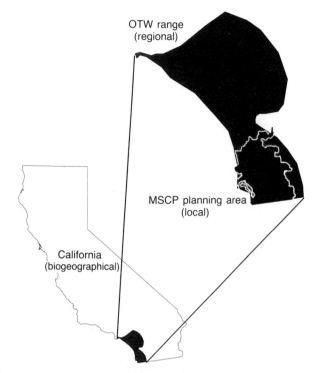

FIGURE 5.3 Range limits of the orange-throated whiptail. The range outline used in this analysis is plotted in relation to California. Also plotted is the region of the MSCP planning area.

of these characteristics becomes important during the interpretation of the species distribution data.

5.3.2 Whiptail observation data sets

Four datasets are direct sources of whiptail distribution information for this study. None of these data sets cover the portion of the range in Baja California, so we restrict the analysis to the section of the range in the state of California. The coarsest representation, used for the biogeographic extent, is a range map digitized from a state map at the 1:3 500 000 scale (Zeiner *et al.*, 1990). The second data set consists of locality information for 349 museum specimens of whiptails. This data set was compiled by Mark Jennings in a thorough

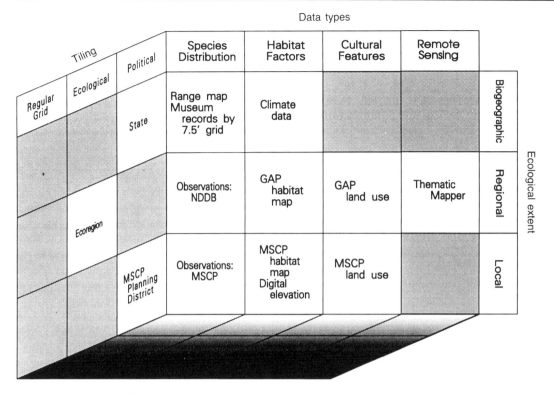

FIGURE 5.4 Biodiversity hypercube filled in with the orange-throated whiptail data sets used in this analysis. All data sets represent a single point on the time dimension which is not shown.

survey of major museum collections. Since geocoding museum locality information presents many difficulties (McGranaghan and Wester, 1988), we positioned each of the locality records in the nearest US Geological Survey 7.5′ quadrangle. As this is a coarse spatial grain covering the entire northern range, we therefore treat this data set as the biogeographic extent.

The third data set consists of 61 point observations for the whiptail from the California Natural Diversity Database (CNDDB). CNDDB is maintained and compiled by the Natural Heritage Division of the California Department of Fish and Game, and it currently has about 19 000 observations, both from museum collections and field observations, of rare and sensitive plant and animal species. Since the whiptail data are concentrated in the more coastal,

urbanized portion of the range, we consider the CNDDB data set as the regional extent. The last data set covers the planning area of the Multiple Species Conservation Plan (MSCP) in western San Diego County, and includes 432 sightings of orange-throated whiptails compiled by OGDEN Environmental and Energy Services. This data set is used as the local extent. The hypercube in Figure 5.4 is filled in with these concrete examples of data sets. This figure represents the relations between all the data used in this analysis.

5.3.3 Species range mapping: Predicting species range limits

The first step in developing representations of a species distribution is to predict the overall range limits of the species.

The range limits considered in this study (Figure 5.3) were superimposed onto the available 7.5' quadrangles (Figure 5.5). The shaded quadrangles contain sightings from the geocoded museum records, the MSCP data set, and the CNDDB data set. The year of the most recent observation is indicated. Several points are evident from this overlay (Figure 5.5). There is reasonable agreement between the range outline and the quadrangle representation. Only the spur at the northwest corner of the range outline appears to be incompatible with the quadrangle layer. Also, the sampling intensity appears to influence the degree of filling of the range outline. The southern portion of the map, with the extensive MSCP sampling, is clearly distinct from the northern portion with older, scattered observations. Thus it is not possible to determine whether the absence of filled quadrangles along the northern coast is a result of a lack of data records or that the species is absent from this area. Finally, the county outlines in Figure 5.5 hint at loss of detail that occurs when distributions are represented by only county of occurrence.

Another approach to capturing the limits to a species' distribution is to relate these boundaries to climatic conditions. The limits to distributions can be set on the biogeographic extent because organisms are physiologically adapted to a hyperspace of climatic conditions. This approach is termed 'bioclimatic modeling', and it has been developed particularly by Australian scientists (e.g. Lindenmayer *et al.*, 1991; Nix, 1989), although Root (1988) also applied this approach using a less automated technique. The link between climate and distribution limits is not direct but is mediated by other biotic and abiotic interactions. For example, Porter and Tracy (1983) suggested that the distribution limits for the desert iguana (*Dipsosaurus dorsalis*) are set not by the thermal environment directly controlling adult distribution but by its effects on soil moisture which strongly affects egg development. Nevertheless,

a climatic representation of a distribution may be useful for applications such as models of climatic gradients and global climatic change (Lindenmayer *et al.*, 1991).

In this study we take the existing range outline and evaluate its correspondence with regional climatic patterns. First we overlaid the range outline on a raster data set of various climatic parameters to develop a statistical model of the association between the range and climate. This data set is from the ZedX raster data set (ZedX, 1991) of climatic information for California, and consists of five data layers at 30 arc-second (about 1 km^2) resolution. These data layers are the mean temperature of the three warmest months, the mean temperature of the three coldest months, the mean annual temperature range, annual precipitation and a derived moisture index. For this study a 1% random sample was taken of pixels falling within a rectangular window surrounding the range outline. The rectangular window buffered the range outline by at least 15 km. This size was chosen to equalize the sampled area falling inside and outside the range outline.

We developed a hierarchical model based on climatic variables to predict whether pixels fell inside or outside the range outline. Such a classification tree recursively partitions the independent variables according to the strength of relationships between each independent variable and the dependent variable (Breiman *et al.*, 1984). The size of this tree is determined subjectively based on a measure that weighs the cross-validated predictive power of a given split against the complexity that the split adds to the tree.

This classification tree is represented in Figure 5.6. This particular tree contains 10 terminal nodes and has a misclassification rate of 13%. The terminal nodes represent the 10 different classes into which a pixel can be placed. Each terminal node is labeled with the number of sample pixels in each class (out of a total of 34 019 pixels) and whether the majority

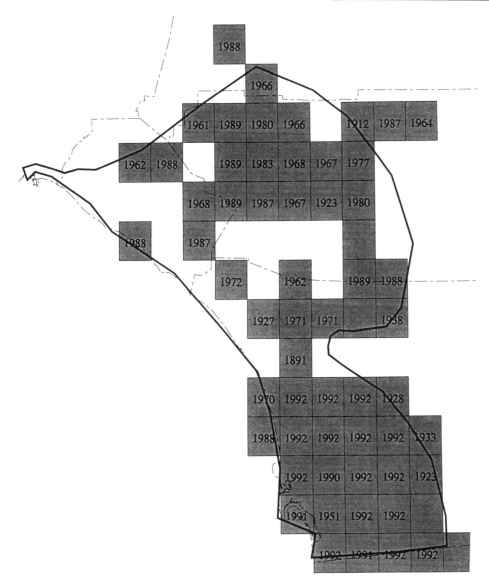

FIGURE 5.5 Whiptail range limits. The generalized range outline is shown in heavy lines; the shaded grid cells are 7.5′ quadrangle cells in which there are museum records, or sightings from either the CNDDB or MSCP database. The date in each cell is that of the most recent sighting, unless there is no information on date of sighting.

of the pixels in the class fall inside or outside the range outline. The nonterminal nodes are labeled with the criterion for choosing which branch of the tree to follow in classifying each pixel. Thus the first split in the tree was whether the minimum temperature variable fell below 8.2°C. Pixels under this threshold (following the left-hand branch) were highly likely to be outside the range outline. Pixels with minimum temperature above this threshold (following the right-hand branch) were about equally as likely to be inside as outside the

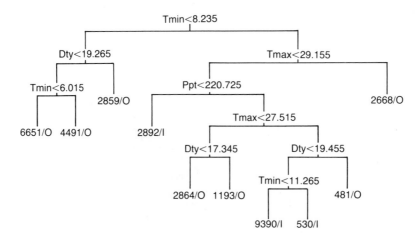

FIGURE 5.6 Classification tree to predict whether a pixel is inside or outside of the range outline. The four independent variables represented are mean temperature of the three warmest months (Tmax), mean temperature of the three coldest months (Tmin), annual temperature range (Dty), and annual precipitation (Ppt). One additional variable, a moisture index, does not appear in this tree as it has been pruned back to 10 terminal nodes. Indicated on each nonterminal node of the tree is the decision rule applied at that branch – e.g. if less than the number indicated in the criterion, the left-hand branch is taken. At the terminal nodes it is indicated whether the branch predicts a pixel to be inside (I) or outside (O) the range outline together with the total number of sample pixels (out of 34 019) applying to that class.

range outline. The next most important distinguishing variables were maximum temperature and annual precipitation.

The classification tree provides a way to categorize any pixel within the study area to a single climate class based upon the values of the climate variables. All pixels contained within the data 'window' defined by the study area are categorized using this approach. Figure 5.7 is a probabilistic representation of the whiptail distribution. Higher values are presented in Figure 5.7 with darker pixel shades. This figure shows a relatively good correspondence between this classification and the species range outline along the eastern portion of the range. This suggests that the minimum temperature boundary corresponds to a limiting factor on the whiptail's distribution in this area. Along the northern boundary of the range, there is little agreement between the classified image and the range outline. Therefore climate appears to be an unimportant factor limiting the distribution of the species in this area. This corresponds

with the area north of the range outline that lies in the unsuitable habitat of a major metropolitan area. The relation between climate and species distribution is thereby identified in the study area using this classification system.

5.3.4 Mapping suitable habitats

(a) Regional habitat mapping

Once the range limits are defined, the next refinement is to exclude unsuitable habitat areas within the range outline. Vegetation is used often as the most easily accessible variable that facilitates regional habitat suitability mapping. Here we relate a vegetation map to a habitat relationships model to predict the distribution of the whiptail within the generalized range boundary. This is the same approach taken by the US Fish and Wildlife Service Gap Analysis Program (Scott *et al.*, 1993). The habitat model used here is the California Wildlife Habitat Relationships (WHR) model

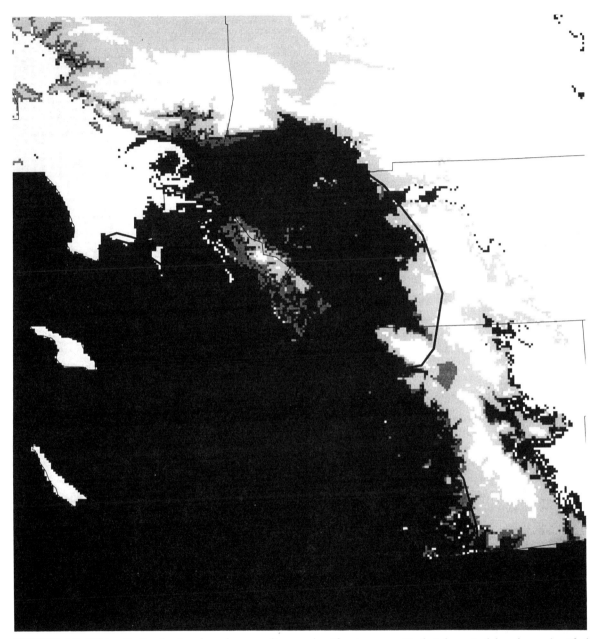

FIGURE 5.7 Whiptail distribution based upon climate classification tree. Each 1-km pixel has been classified based on the tree in Figure 5.6, and is shaded proportionally to the percentage of sampled pixels in that class falling within the range outline. The generalized range outline corresponds fairly well to a break in the classified image on its eastern boundary, but less so on the northern boundary.

(Airola, 1988), a tabular database with information on habitat preferences for 646 species of terrestrial vertebrates resident within the state. This database categorizes wildlife habitat into 48 major vegetation types that are further subdivided into vegetation structural categories. In this study we only consider the major vegetation types. These habitat types are mapped over the study area by the California Gap Analysis Program. These maps are produced with a minimum 100 hectare mapping unit (Davis and Stoms, 1992) at a 1 : 100 000 scale using digital Landsat Thematic Mapper (TM) imagery, aerial photography and existing maps. Each polygon is predicted to include the whiptail if it contains suitable habitat types defined by the WHR model and if it falls within the generalized species range boundary.

Plate 4 is a predicted distribution for the orange-throated whiptail produced using the above approach. Suitable WHR habitat types used in this prediction include mixed chaparral, chamise-redshank chaparral, coastal sage scrub and valley-foothill hardwood. Plate 4 also includes the generalized range boundary for this species. This original range outline has an area of 11 300 km^2, in contrast to the WHR distribution model that includes an area of only 6900 km^2. This is to be expected because not all habitats within the range outline are suitable.

There was some discordance between the location of the 61 point observations from CNDDB – also plotted on Plate 4 – and the polygons in the WHR model. This discrepancy could be due to many factors including a scale mismatch between the point observations and the WHR model, a difference in the time of data collection, or inadequacies in the WHR model. However, an insufficient number of points are available to assess which of these factors is the likely cause of this problem. In the next section, a more localized data set permits a more detailed analysis of the mapped areas.

(b) Local habitat mapping

Local areas often hold data sets with a greater quantity and detail of information than regional areas. In this study, we examine a

detailed habitat map of the southern portion of the whiptail range in California, a compilation of recent orange-throated whiptail observations in this area, and a Landsat TM image for this region. The goal of this portion of the study was to extrapolate relationships over a broader region from the local data.

The large-scale vegetation map used in this analysis is of the MSCP planning area in western San Diego County. It was compiled from an overlay of color infrared aerial photos (June 1990) onto a 1:24 000 scale base by biologists at OGDEN Environmental and Energy Services. Vegetation boundaries were field checked with helicopter overflights. The map was reviewed and updated with field surveys in 1992. Polygons were labeled using the 'Holland' classification system for natural communities (Holland, 1986). This map permitted us to distinguish disturbed and undisturbed communities.

As previously described, the 432 orange-throated whiptail sightings were compiled from many sources. The data set is not a random sample. The projects for which whiptail sightings were collected required environmental impact reports (EIRs) which are usually for urban development, water development or road construction. Therefore the observations were taken primarily from the urban fringe, where most development occurs, and this pattern is evident through visual inspection of the data.

To identify the degree of this collecting bias on the distribution of observations, the distance of each observation from the nearest developed area was derived through standard GIS operations. Observations included within polygons containing developed areas were not considered in this analysis. Precise distances were aggregated into classes for analysis of the distribution. Of the 421 sightings that were not in developed areas, 408 of these were within 2 km of a developed area boundary, and 343 of these sightings were within 1 km of a developed area. Over 50% of the observations were within 500 m of development. The mean distance of the sightings from a developed area was 550 m, with a standard deviation of 532 m. By contrast, the mean distance of all suitable habitat pixels on the MSCP vegetation map from developed areas was 3711 m, with a standard deviation of 1466 m. Therefore, even for an intensively observed species, the data are usually located in areas that are planned for development in the near future. Consequently, these observations are more frequently located in habitats at risk rather than habitats ecologically favored by the species.

Another source of bias in the whiptail sightings is that the MSCP data set does not sample the full elevational range of the species. The species notes published with the WHR system (Zeiner *et al.*, 1990) indicate that the elevational distribution of the orange-throated whiptail is less than 900 m. When the whiptail observations were superimposed upon a digital elevation model (DEM) with a 100 m resolution grid, all observations in the MSCP data set were below 800 m and 88% were below 400 m in elevation. This may once again reflect the greater number of observations taken from urban fringe areas, but it also demonstrates that most of the MSCP area is less than 500 m in elevation and therefore does not represent potential whiptail habitats at higher elevations.

The habitat map for the whiptail was generated using the same logic as that used with the WHR regional map. The Holland communities were reclassified into WHR habitat types using the crosswalk presented in Mayer and Laudenslayer (1988). These habitat types were then aggregated into the three classes of suitable habitat (i.e. coastal sage scrub, mixed chaparral, chamise-redshank chaparral and valley-foothill hardwood), disturbed habitat (same four habitat types above, but indicated to be disturbed on the source map), and unsuitable (i.e. all other habitat types). Using the GIS, points were overlaid with the recoded habitat map, and summary statistics were produced for

polygons with and without sightings. Summary statistics were compiled for the three suitability categories for total area, average polygon size and median polygon size. This representation is plotted in Figure 5.8.

'Suitable' habitat polygons contained observations much more frequently than 'unsuitable' habitat. This was indicated through relatively few errors of commission (roughly 79% accuracy by polygon area considering suitable (62%) or disturbed (17%) habitats). However, whiptails have not been recorded in polygons representing 73% of the area predicted by the WHR model as suitable or disturbed habitat. In the absence of information about where biologists sampled but did not observe this species, the 'empty' habitat polygons cannot be considered errors of omission. Rather, they suggest locations for additional sampling.

An analysis of the location of observations within habitat types on the MSCP map demonstrates general agreement with WHR suitability ratings, but no observations are recorded in valley-foothill hardwoods, though the WHR model lists this as suitable whiptail habitat. Chaparral is also shown as suitable habitat using WHR, and although 71 observations (out of 432) are recorded in chaparral, this was fewer than expected from this habitat's availability. Surprisingly, there are a high number of sightings (67 observations) shown by whiptails in disturbed coastal scrub habitat. It is unclear whether this is accurately interpreted as indifference to habitat disturbance, or whether this could be explained by degradation of the habitat since the date of the observations.

(c) Use of high-resolution satellite imagery for local habitat mapping

Satellite imagery is a useful adjunct to thematic habitat maps since raster modeling is more practical for many purposes and allows extrapolation over a larger extent. Spectral classes from satellite imagery usually do not correspond to information classes of thematic maps.

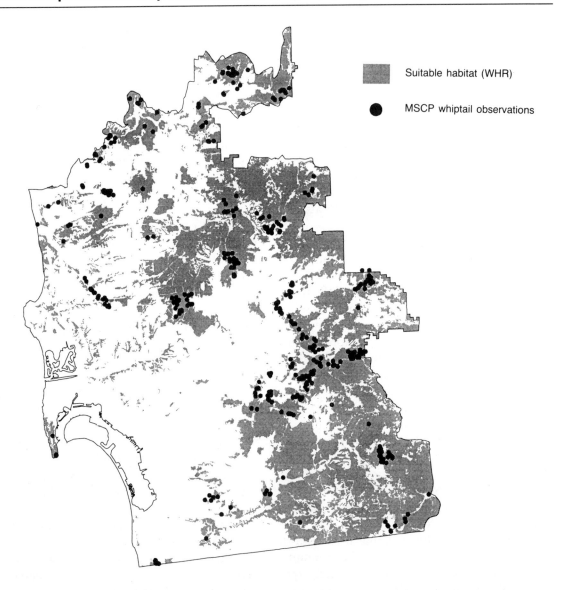

Suitable habitat (WHR)

MSCP whiptail observations

FIGURE 5.8 Whiptail habitat and observations within the MSCP study area. Shaded areas are suitable habitat according to the WHR model mapped onto the MSCP vegetation map; dots represent the whiptail sightings within the MSCP database. Most observations fall on the edge of the urban area.

However, spectral classes can be used to categorize wildlife habitats across the landscape in useful ways.

To test whether satellite and digital terrain data could be used to refine the range map of the whiptail, we compared the field data with a classified TM image. An unsupervised classifi-

cation was performed on the red, near-infrared and mid-infrared bands (3, 4 and 5 respectively) for four 1990 TM scenes, with digital elevation data added as a fourth channel. Forty unlabeled clusters were created after we masked urban and agricultural areas. The MSCP observation data were then set within a grid with the same

100 m resolution as the satellite imagery and these data were superimposed onto the classified image. The observations were most frequently located within three of the 40 clusters. These three clusters were subsequently considered 'suitable' habitat for the purposes of this analysis. This correspondence was considered strong since 63% of the whiptail sightings fell within these three clusters, even though these clusters occupied only 13% of the area of the ecoregion.

One use for a raster habitat representation is for the analysis of properties such as patch contiguity and fragmentation. Such patterns are of much concern to conservation biologists (e.g. Rolstad, 1991). To test the efficacy of such an analysis in our study we extracted these clusters from the grid and overlaid them onto the MSCP vegetation polygons with positive associations to whiptail observations (coastal sage scrub and oak riparian). A contiguity value was computed for each pixel in the area of intersection and a contiguity index was subsequently derived for each of the intersected habitat patches (LaGro, 1991). This contiguity index, derived from satellite imagery, can further discriminate ecological conditions beyond simple habitat types, to the degree that a lack of sightings can be interpreted as an indication of unsuitable or unoccupied habitat. Low contiguity scores (approaching 1.0) indicate small and isolated patches whereas contiguity scores approaching 2.0 indicate large polygons containing many contiguous pixels of the three preferred spectral clusters. Polygons with whiptail sightings had a higher mean contiguity index (1.76) than polygons without sightings (a value of 1.54). This suggests that properties such as patch contiguity in this species deserve further investigation.

Habitat maps derived from a spectral classification also permit extrapolation to some larger spatial extent. Here we plot (Plate 4) spectral classes across the whole range outline of the whiptail. This provides a useful summary of the predicted occurrence of the orange-throated whiptail distribution based upon these three distinct sources of information, i.e. the range boundary, the habitat map and the multispectral data. The entire range boundary delineates an area encompassing 11 300 km^2. The habitat map, based upon suitable WHR habitat types, encompasses an area of 6900 km^2. The frequently used spectral classes, when extrapolated for the entire range, total only 2200 km^2. These three data sources provide quite different areal extents for the whiptail distribution.

5.4 Discussion and conclusions

This whiptail study illustrates how different distribution and environmental data at various scales can generate predictive distribution maps and hypotheses about the factors controlling them. None of these representations can be considered definitive, but each has its uses. The range outline is a coarse-scale representation of a distribution, but is adequate for analyses at the biogeographic scale. These might include assessment of species richness patterns at a subcontinental scale (Kiester, 1971). A species distribution presented as a grid cell representation based upon collection data (e.g. which preserves data certainty) together with the range outline indicates poorly studied areas which require further survey. A habitat-based model circumvents the problem of inadequate and biased data collection, but calls for the habitat requirements to be well known. Typically, this is not the case, as the habitat requirements are only generally known. Point observation data can be superimposed onto the habitat-based model to show areas of inadequate sampling. Finally, satellite imagery shows a synoptic view of the landscape and provides a useful alternative characterization of the potential habitat. The advantages of each map approach become more apparent when these different representations are considered together.

This set of representations for the orange-throated whiptail suggests several additional lines of investigation – for example, to make further samples in poorly studied areas. The areas of suitable WHR habitat in Plate 4 both in the northwest and northeast portions of the range outline are candidates for additional survey. Further sampling would allow for the iterative refinement of the habitat models, which would include spectral models (Skidmore, 1989). The further testing of habitat models is another line of investigation suggested by these results. The WHR database predicts the whiptail to occur within more sparsely vegetated shrublands, but we did not attempt to include vegetation crown cover as a predictive factor of the whiptail distribution. However, vegetation crown cover can be feasibly mapped with satellite imagery (Franklin *et al.*, 1986), and testing the preference of whiptails for vegetation of different crown closures can be demonstrated using a sampling scheme devised using a GIS (e.g. Davis *et al.*, 1992).

To refer back to the hypercube, this three-tiered scheme illustrates the importance of casting the diversity of data into a suitable structure. Within this structure, as the whiptail example has shown, different data sets are compared looking for common patterns and making inferences. Effectively this process constitutes a sensitivity analysis on the predicted distribution for the species. Such an interactive process can easily be extended to more general problems in conservation planning.

Unfortunately, existing commercial GISs do not facilitate this sort of interactive work. The heterogeneity of these data sets – composed as they are of vector maps, images, tabular and statistical models, and so on – means that much effort must be put into converting data from one form to another. GISs are weak in enabling spatial statistical analyses, and systems that assist in searching for geographical patterns are

only at the research stage (Goodchild *et al.*, 1992; Openshaw *et al.*, 1990); but, conversely, improving ability to visualize and integrate complex data sets is an active area of development and research in GIS (Shepherd, 1991).

A promising approach for such integration has come from declarative or logic programming languages that have emerged from the artificial intelligence paradigm. These languages, if linked to a database, can be viewed as powerful extensions to the relational database model that is standard to many GISs, providing inference capabilities and more expressive representations of knowledge (Smith and Yiang, 1991; Egenhofer and Frank, 1990). Rule-based reasoning systems supported by these languages can be valuable assets to complex resource management tasks (Loh and Rykiel, 1992). Moreover, the rigor automation enforces on expressing models in logic makes evident their structure and assumptions (Muetzelfeldt *et al.*, 1989). Many of the models used in conservation planning are logical ones (e.g. a plant species is predicted to be on a site if it is below 500 m elevation, is on clay soils and is on a flat or gentle slope). Indeed, such logical formulations of models may be more appropriate for conservation planning than the gradient models traditional to vegetation science (Davey and Stockwell, 1991).

Thus we envision a mapping environment where the researcher no longer struggles to produce a single map, but produces suites of them at will. Data integration is one component to this, but so are the flexibility and clarity of the underlying models – the multiple images thereby creating a better representation of the complex reality underlying diverse data sources.

5.5 Summary

Geographic Information Systems technology permits the generation of complex representations of species distributions, while most of the data underlying these patterns are coarse. This suggests the importance of structuring such data along axes of differing data extent, tiling schemes, themes and time, and displaying different representations of distributions, the philosophy being that comparison of multiple representations provides a sense of the actual distribution through convergence of evidence. We present an example using a lizard, the orange-throated whiptail (*Cnemidophorus hyperythrus*), which is native to southern California. The analysis was hierarchically structured by first mapping overall lizard range limits, then suitable habitats within the range, and then habitats over a local extent. Data sources include a generalized range outline, museum records and field observations, as well as climate data, vegetation maps and satellite imagery to serve as associated environmental variables. Comparison of representations resulting from these different data sources makes biases evident, highlights areas of inadequate sampling and can lead to new inferences about habitat relationships. Finally, we discuss forthcoming improvements in the technology that will facilitate creation and display of families of models.

Acknowledgements

Many people graciously provided the datasets used in this analysis, including Mark Jennings, Pete Stine, Janine Stenback, Tom Lupo, Reg Barrett and Robert Motroni. We would also like to thank Steve Sherrill and Mike Bueno for data entry and processing. Funding for this work has been provided by the National Fish and Wildlife Foundation and Southern California Edison. Additional support was provided by the IBM Environmental Research Program. The manuscript was reviewed by James Quinn and Barrett Garrison.

References

Airola, D.A. (1988) *Guide to the California Wildlife Habitat Relationships System*, State of California, The Resources Agency, Department of Fish and Game, Sacramento, California, 74pp.

Bailey, R.G. (1989) Explanatory supplement to eco-regions map of the continents. *Environmental Conservation*, **16**, 307–9.

Berry, B.J.L. (1964) Approaches to regional analysis: a synthesis. *Annals of the Association of American Geographers*, **54**, 2–11.

Bostic, D.L. (1965) The ecology and behavior of *Cnemidophorus hyperythrus beldingi* Cope (Sauria: Teiidae). Unpublished Master's thesis, San Diego State College, San Diego, California.

Bostic, D.L. (1966) Thermoregulation and hibernation of the lizard, *Cnemidophorus hyperythrus beldingi* (Sauria: Teiidae). *Southwestern Naturalist*, **11**, 275–89.

Breiman, L., Friedman, J.H., Olshen, R.A. and Stone, C.J. (1984) *Classification and Regression Trees*, Wadsworth, Belmont, California, 358pp.

Daubenmire, R.F. (1978) *Plant Geography: with Special Reference to North America*, Academic Press, New York, 338pp.

Davey, S.M. and Stockwell, D.R.B. (1991) Incorporating wildlife habitat into an AI environment: concepts, theory, and practicalities. *AI Applications*, **5**, 59–104.

Davis, F.W. and Stoms, D.M. (1992) Gap analysis of biodiversity in California, *Proceedings of the Symposium on Biodiversity of Northwestern California, October 28–30, 1991, Santa Rosa, California*. University of California Wildland Resources Center, Berkeley, California, pp. 23–9.

Davis, F.W., Stoms, D.M., Estes, J.E. *et al.* (1990) An information systems approach to the preservation of biological diversity. *International Journal of Geographical Information Systems*, **4**, 55–78.

Davis, F.W., Schimel, D.S., Friedl, M.A. *et al.* (1992) Covariance of biophysical data with digital topographic and land use maps over the FIFE site. *Journal of Geophysical Research*, **97**, 19009–21.

Egenhofer, M.J. and Frank, A.U. (1990) LOBSTER: Combining AI and database techniques for GIS. *Photogrammetric Engineering and Remote Sensing*, **56**, 919–26.

Folse, L.J., Packard, J.M. and Grant, W.E. (1989) AI modelling of animal movements in a heterogeneous habitat. *Ecological Modelling*, **46**, 57–72.

Franklin, J., Logan, T.L., Woodcock, C.E. and Strahler, A.H. (1986) Coniferous forest classification and inventory using Landsat and digital terrain data. *I.E.E.E. Transactions, Geoscience and Remote Sensing*, GE-24:139–149.

Goodchild, M. Haining, R., Wise, S. *et al.* (1992) Integrating GIS and spatial data analysis – problems and possibilities. *International Journal of Geographical Information Systems*, 6, 407–23.

Hanski, I. and Gilpin, M. (1991) Metapopulation dynamics – brief history and conceptual domain. *Biological Journal of the Linnean Society*, 42, 3–16.

Holland, R.F. (1986) *Preliminary Descriptions of the Terrestrial Natural Communities of California.* State of California, The Resources Agency, Nongame Heritage Program, Department of Fish and Game, Sacramento, California, 156pp.

Kiester, A.R. (1971) Species density of North American amphibians and reptiles. *Systematic Zoology*, 20, 127–37.

LaGro, J., Jr (1991) Assessing patch shape in landscape mosaics. *Photogrammetric Engineering and Remote Sensing*, 57, 285–93.

Levin, S.A (1992) The problem of pattern and scale in ecology. *Ecology*, 73, 1943–67.

Lindenmayer, D.B., Nix, H.A., McMahon, J.P. *et al.* (1991) The conservation of Leadbeater's possum, *Gymnobelideus leadbeateri* (McCoy); a case study of the use of bioclimatic modelling. *Journal of Biogeography*, 18, 371–83.

Loh, D.K. and Rykiel, E.J. (1992) Integrated resource management systems – coupling expert systems with data-base management and geographic information systems. *Environmental Management*, 16, 167–77.

Mayer, K.E. and Laudenslayer, W.F. Jr (1988). *A Guide to Wildlife Habitats of California.* State of California, The Resources Agency, Department of Fish and Game, Sacramento, California, 166pp.

McGranaghan, M. and Wester, L. (1988) Prototyping an herbarium collection mapping system. *Technical Papers: 1988 ACSM-ASPRS Annual Convention: GIS*, 5, 232–8.

Muetzelfeldt, R., Robertson, D., Bundy, A. and Uschold, M. (1989) The use of Prolog for improving the rigour and accessibility of ecological modelling. *Ecological Modelling*, 46, 9–34.

Nix, H. (1989) A biogeographic analysis of Australian elapid snakes, in *Atlas of Elapid Snakes of Australia* (ed. Richard Longmore). Australian Flora and Fauna Series, No. 7. AGPS Press, Canberra, pp. 4–15.

Openshaw, S., Cross, A. and Charlton, M. (1990) Building a prototype Geographical Correlates Exploration Machine. *International Journal of Geographical Information Systems* 4, 297–311.

Parker, I. and Matyas, W.J. (1981) *CALVEG: A Classification of Californian Vegetation*, US Dept. of Agriculture, US Forest Service, Regional Ecology Group, San Francisco, 168pp.

Porter, W.P. and Tracy, C.R. (1983) Biophysical analyses of energetics, time-space utilization, and distributional limits, in *Lizard Ecology* (ed. R.B. Huey, E.R. Pianka and T.W. Schoener), Harvard University Press, Cambridge, Mass., pp. 55–83.

Rapoport, E.H. (1982), *Areography: Geographical Strategies of Species*, Pergamon Press, Oxford, 269pp.

Rolstad, J. (1991) Consequences of forest fragmentation for the dynamics of bird populations – conceptual issues and the evidence. *Biological Journal of the Linnean Society*, 42, 149–63.

Root, T.L. (1988) *Atlas of Wintering North American Birds: An Analysis of Christmas Bird Count Data*, University of Chicago Press, Chicago, 312pp.

Scott, J.M., Davis, F., Csuti, B. *et al.* (1993) Gap analysis: A geographic approach to protection of biological diversity. *Wildlife Monographs*, 123, 1–41.

Shepherd, I.D.H. (1991) Information integration and GIS, in *Geographical Information Systems*, Vol. 1: *Principles* (ed. D.J. Maguire, M.F. Goodchild and D.W. Rhind), Longman, Harlow, Essex, England, pp. 337–60.

Shimwell, D.W. (1971) *The Description and Classification of Vegetation*, University of Washington Press, Seattle, 322pp.

Skidmore, A.K. (1989) An expert system classifies eucalypt forest types using Thematic Mapper data and a digital terrain model. *Photogrammetric Engineering and Remote Sensing*, 55, 1449–64.

Smith, T.R. and Yiang, J. (1991) Knowledge-based approaches in GIS, in *Geographical Information Systems*, Vol. 1: *Principles* (ed. D.J. Maguire, M.F. Goodchild and D.W. Rhind), Longman, Harlow, Essex, England, pp. 413–25.

Stebbins, R.C. (1985) *A Field Guide to Western Reptiles and Amphibians*, Houghton Mifflin, Boston, 336pp.

Turner, M.G., Dale, V.H. and Gardner, R.H. (1989) Predicting across scales: theory development and testing. *Landscape Ecology*, 3, 245–52.

Walker, P.A. (1990) Modelling wildlife distributions using a geographic information system – kangaroos in relation to climate. *Journal of Biogeography*, 17, 279–89.

Wiens, J.A. (1989) Spatial scaling in ecology. *Functional Ecology*, 3, 385–97.

Woodward, F.I. (1987) *Climate and Plant Distribution*, Cambridge University Press, New York, 174pp.

ZedX (1991) *Hi-Rez Data: Climatological series – California*, ZedX Inc., Pennsylvania.

Zeiner, D.C., Laudenslayer, W.F. Jr, Mayer, K.E. and White, M. (ed) (1990) *California's Wildlife*, 3 vols, State of California, The Resources Agency, Department of Fish and Game, Sacramento, California.

Part Four

Mapping Migratory Species Distribution Patterns

Chapter 6 presents details of contemporary ornithological field survey methods that are used in tropical forests and it shows how these data are then linked to satellite imagery to produce maps. These methods are used to assess habitat availability for the wood thrush, *Hylocichla mustelina*, a nearctic migrant bird species that winters in an area that has undergone major deforestation in recent decades — the Caribbean lowlands of Middle America. Landsat imagery is used to identify vegetation types in northeastern Costa Rica and southern Belize and previously collected data are used to identify vegetation types in southern Veracruz,

Mexico. Field surveys of random locations in each country are then conducted to validate these vegetation maps. On sites representing each of the major vegetation classes, mist-net and audio-visual censuses were performed. These approaches were used to identify the winter range habitats where wood thrushes were most likely to occur. This chapter clearly demonstrates that valuable estimates of habitat distribution and rates of change can be derived from satellite imagery combined with extensive ground censuses. These estimates are extremely useful for determining the conservation status of a nearctic migrant.

Chapter six

Remote-sensing assessment of tropical habitat availability for a nearctic migrant: The wood thrush

John H. Rappole,
George V.N. Powell and
Steven A. Sader

6.1 Introduction

The wood thrush (*Hylocichla mustelina*) is a common understorey species with a range that varies along with the season of the year. During the breeding season it inhabits deciduous and mixed woodlands (Phillips, 1991). During the north temperate winter months (i.e. November–March) the species is found in the wet lowlands of Middle America from southern Mexico to Panama (Rappole *et al.*, 1983; Phillips, 1991) (Figure 6.1).

According to Breeding Bird Survey data (Robbins *et al.*, 1989), population declines are evident in recent years for the wood thrush and many other nearctic avian migrant species with forest habitat preferences and winter ranges in the neotropics. Breeding habitat did not decline, but severe losses in winter habitats did occur over this period (DeGraaf and Rappole, in press). These population declines heighten debate about which portion of the annual cycle is most critical to survivorship, and this

Mapping the Diversity of Nature. Edited by Ronald I. Miller.
Published in 1994 by Chapman & Hall, London. ISBN 0 412 45510 2.

FIGURE 6.1 The wood thrush range observed in this study.

prompts questions about where declines occur and the nature of their causal factors.

The problem of conservation of migratory species with complex life cycles is difficult. Species such as the wood thrush that fail to 'stay put' throughout the course of an annual cycle often are viewed as unlikely candidates for conservation action. The apparent reasoning is that migratory organisms are niche generalists and are thus able to exploit a wide variety of habitats.

This concept is well articulated by Morse (1971: 186): 'The problem of utilizing several habitats within a year should result in migrants exhibiting greater plasticity and being less specialized than residents.' Results from several subsequent field studies support this hypothesis, and indicate that many migratory species use a wide range of habitats while on their wintering grounds (Karr, 1976; Hutto, 1980; Petit *et al.*, 1992).

An alternative theory is that migratory species are dependent on the several different habitats that they use at different latitudes, and the loss of any one of these could cause their extirpation (Rappole *et al.*, 1983). Certainly, a migratory species will not persist if its principal breeding habitat is destroyed. Are stopover or winter habitats any less critical?

As a result of environmental destruction, these opposing hypotheses are being tested globally. Neotropical forests are disappearing at a rapid rate (Sader and Joyce, 1988; Dirzo and Garcia, 1992; World Resources Institute, 1992), and 107 species of migratory birds are known to use these forests as their principal wintering habitat (Rappole *et al.*, 1983). The continued survival of a number of migratory species may depend upon the degree to which conservation action can mitigate the destruction of these habitats.

6.2 Study context

Remote-sensing technology provides a feasible way to examine the abundance of remaining habitat and the influence of habitat change on migratory bird populations (Green *et al.*, 1987; Sader *et al.*, 1991; Powell *et al.*, 1992). In this chapter, we analyze vegetation cover information from satellite imagery and on-site sampling data in relation to wood thrush winter habitat-use data. Our objectives are: (1) to determine if remote-sensing technology can be used to identify wood thrush habitats; (2) to develop density indices for the wood thrush that are suitable for use in different habitat types; and (3) to test these techniques in different parts of the winter range. We subsequently use these data to predict the consequences of the current trends in habitat change in Middle America on the future of the wood thrush in this region.

6.3 Methodology

6.3.1 Study areas

We collected data during several winter field seasons in three Middle American countries:

Northeastern Costa Rica
    Provinces: Heredia
                   Cartaga
                   Alejuela
                   Limon
    9 November–4 December 1987
    3–30 November 1988
    13–22 February 1989
Southern Belize
    District: Toledo
    3–18 November 1989
    5–20 February 1990
    5 November 1990–28 February 1991
Southeastern Mexico
    State: Veracruz
    District: Catemaco
    November 1980–4, 1986

The Mexican data are from two lowland rainforest sites in the Tuxtla Mountains of southern Veracruz; the Biology Station at the University of Mexico (November 1980–82), and a study area near the village of La Peninsula de Moreno, 15 km east from the town of Catemaco (November 1986).

6.3.2 Remote sensing

Landsat Thematic Mapper satellite scenes (each covering 34 225 km^2 approximating a square 185 km on a side) were purchased for regions in Costa Rica and Belize. We limit our analysis to the wet life zones (Tosi, 1969), i.e. regions of Middle America covered originally with evergreen broadleaf forest. Both satellite scenes are from approximately two years prior to our field surveys: Costa Rica, 6 February 1987; Belize,

25 March 1987. Only a portion of each scene is selected for analysis during this study. In Costa Rica we select a subscene approximately equal to 2016 km^2 and in Belize our subscene covers 810 km^2 (see Plates 5 and 6 respectively. In Plate 6 the subscene area is reduced to minimize problems with cloud cover and haze contamination in the imagery.

Subscenes are processed initially to provide an unsupervised classification. The satellite image is analyzed statistically to correlate the reflectance data with the various ground cover types in each scene. The Landsat data are recorded by the Thematic Mapper (TM) sensor for which each cell or pixel corresponds to a ground quadrat equal to 30 m on a side. Pixel clusters are created based upon the similarity of multispectral reflectance characteristics. Further groupings are developed from a correspondence between pixel locations and the ground locations of specific land cover types (determined from analysis of aerial photos). Habitat groups are formulated using this approach, and each is assigned to a unique map color (Plates 5 and 6).

6.3.3 Ground truth of satellite data

We select random points for the ground survey to determine the accuracy of the groupings. In Costa Rica, 142 sites are examined and 80 sites are surveyed in Belize. For each site, we record a physical description of the habitat, photograph the site, and identify each site with the appropriate TM image classification. Based on the site surveys, we then compress all the habitats into the following three major categories based on the predominant foliage height: (1) forest >10 m, (2) second growth 3–10 m, and (3) open areas <3 m. We then select random points within each of these major habitat groupings for an analysis of migratory bird numbers.

6.3.4 Mist-net sampling

Migratory bird numbers are measured by mist-net sampling and by audio-visual census. The mist-net procedure in Costa Rica and Belize involves placement of a grid comprising 13 nets onto each randomly selected site (1 ha). Each grid consists of five rows, 100 m in length, and each row is separated by 25 m. Each net consists of 36 mm mesh, and each extends 12 × 2.6 m. Each grid is sampled for 250 net-hours (1 net-hour = one net open for 1 hour), with two netting days per site, dependent upon the weather. The sex, age, molt condition and fat content are recorded for captive birds, and subsequently each is banded and released. In Costa Rica, we use mist-nets in 13 forest sites (three of these in swamp forest), five second-growth sites, and five pasture sites. In Belize we use mist-nets in 10 forest sites, 10 second-growth sites, and 10 pasture sites.

The Mexico netting data are derived from our earlier studies in the lowland rainforest of that country. These samples are taken at 10 lowland forest sites on 1-ha grids. Each grid comprises 8–10 nets that are placed in three rows, each row is 100 m in length, and each row is separated by 50 m. Each grid is sampled for 250 net-hours. All species counts reported in this chapter from the study sites in each of the three countries are based upon the 250 net-hour survey time.

6.3.5 Audio-visual sampling

We used an audio-visual census method that is a modification of the Variable Circular Plot (VCP) procedure (Reynolds *et al.*, 1980). This method involves visits to 10 points, each 100 m apart, at each randomly selected site. At each location, the observer waits 5 minutes and then records: (1) all individuals seen or heard within 50 m and (2) a distance estimate to the individual.

6.4 Field survey results

6.4.1 Habitat detection derived from the satellite data

Five of the 16 reflectance groups identified in the unsupervised classification are used in our analysis of wood thrush habitat in Costa Rica and eight of these groups are used in Belize (Table 6.1).

As verified during our field surveys, the TM process is quite accurate in identifying major habitat types. Sites identified as 'forest' by the TM imagery (i.e. green or dark brown in the unsupervised classification) are found to be forest or tree crops with a canopy >10 m tall for all but one of the 57 points visited. The exception is an old orange grove with a canopy of 6–8 m (Table 6.1). The TM imagery did not distinguish between disturbed and undisturbed forest, at least not using the classification system employed in this study. Also, older second-growth plots with tall trees are included in the 'forest' color groupings.

Second growth, scrub and crops (orange groves, bananas, shade coffee, cacao, coconuts) with canopy heights of 3–10 m are represented by the colors yellow, purple and red in the unsupervised classification (Table 6.1). Going from red to purple to yellow roughly represents an increase in the canopy height. However, there is considerable overlap in habitat between these groupings.

Pasture habitats (i.e. these include low second-growth habitats and crops with canopies <3 m) are represented by pink, blue and gray in the unsupervised classification. Going from gray to blue to pink also represents an increase in canopy height. As in the case of second-growth habitats, there is considerable overlap in habitat between groupings.

Based on these analyses, we determine that the TM can be used reliably to separate vegetation cover into three meaningful classes for our migratory bird studies: (1) **forest** –

canopy >10 m; (2) **second growth** – canopy 3–10 m; (3) **open** – canopy <3 m. The proportion of each scene covered by these habitats can also be reliably calculated (Table 6.2).

6.4.2 Audio-visual census results

The VCP audio-visual census method was developed originally to provide a measurement of bird species density, based upon estimated distances to birds monitored within a sample circle (radius = 50 m) (Reynolds *et al.*, 1980). We do not use this approach for the wood thrush because detection rates indicate that birds are not recorded on sites where individuals are captured in mist-net samples. Therefore, occurrence is presented as a percentage of the total number of sites surveyed (10 VCP censuses are performed at each site – Table 6.3). Nevertheless, for censusing the wood thrush the VCP approach is inadequate in comparison to mist-net sampling. Using mist-nets in Belize, the wood thrush is found on 90% of both forest and second-growth sites. Using the VCP procedure, the species is detected on 70% of forest sites and only 23% of second-growth sites.

The VCP and comparable audio-visual methods are not particularly useful for censusing quiet, inconspicuous species similar to the wood thrush. Yet these same methods may be useful for species that vocalize a great deal and therefore provide a detectable signal that indicates their presence. These methods also are used to census species that forage in habitats with tall canopies since mist-nets will not provide reliable samples in these environs. Some workers (e.g. Lynch, 1989) use playback of alarm calls (i.e. 'spsshhing') and territorial defense notes to enhance detectability during audio-visual censuses. This procedure may be useful in some circumstances, but we have found that response to playback varies dramatically for different species, and even for different individuals within a species. For example,

TABLE 6.1 A correspondence between habitat types and unsupervised classification of Landsat scenes. The presented correspondences are associated with randomly selected points in Costa Rica and Belize

Habitat types associated with Landsat colors[a]	Costa Rica	Belize
Green		
1. Undisturbed primary forest (>10 m)	3	0
2. Disturbed primary forest (>10 m)	17	6
3. Swamp forest (>10 m)	3	5
4. Tall second-growth forest (>10 m)	2	5
5. Disturbed gallery forest (>10 m)	5	1
6. Old orange grove (6–8 m)	1	0
7. Date palm grove (>10 m)	1	0
Dark brown		
1. Disturbed primary forest (>10 m)	—[b]	3
2. Tall second-growth forest (>10 m)	—	5
Yellow		
1. Second-growth forest (6–10 m)	7	10
2. Tree crops (6–10 m)	14	0
3. Riparian thickets (6–10 m)	5	0
4. Palm swamp (7–9 m)	1	0
Purple		
1. Palm grove (5–8 m)	—	1
2. Pasture with scattered trees (>10 m)	—	1
3. Marsh with scattered shrubs (3–9 m)	—	2
4. Shrubby overgrown pasture (3–5 m)	—	5
5. Bananas (4–6 m)	—	1
Red		
1. Low, shrubby second growth (2–4 m)	3	11
2. Pasture, scattered trees	0	5
3. Shrubby crops	13	0
Pink		
1. Overgrown pasture, fallow fields, marshland (1–3 m)	8	4
2. Low crops (1–3 m)	9	0
3. Grazed pasture, scattered trees	7	0
4. Bananas, pasture, crops mixture	8	0
Blue		
1. Short grass (pasture, marsh) <1 m	15	8
2. Low crops <1 m	7	2
3. Low second growth <1 m	6	2
4. Residential, scattered trees	3	1
5. Mixed crops, pasture	4	0
Gray		
1. Low pasture, marsh <1 m	—	9
2. Soccer field	—	1
Grand total	142	80

[a] See text (6.4.1) for explanation.
[b] — = Colors not represented in the Costa Rica scene.

TABLE 6.2 The proportion of vegetation groups found within the Landsat study sites

Category	Costa Rica	Belize
Forest (>10 m)		
Green	49.8	48.6
Brown	—	24.8
Total forest	49.8	73.4
Second growth (3–10 m)		
Yellow	13.3	6.0
Purple	—	2.3
Red	16.2	1.4
Total second growth	29.5	9.7
Open (<3 m)		
Pink	3.5	5.7
Blue	16.2	2.7
Gray	—	1.1
Total open	19.7	9.5
Other	1.0	7.4
Grand total	100.0	100.0

TABLE 6.3 The percent occurrence of the wood thrush recorded in the Variable Circular Plot surveys on the sites in Costa Rica and Belize

Country	Forest (>10 m)	Second growth (3–10 m)	Open (<3 m)
Costa Rica	60 (10)[a]	0 (2)	0 (12)
Belize	70 (10)	23 (13)	0 (10)

[a] Numbers in parentheses ( ) refer to the total number of sites on which Variable Circular Plot censuses were performed within each habitat type (10 Variable Circular Plot sample circles per site).

response to playback of territorial defense calls in known wood thrush territories on the wintering ground in Veracruz often fails to elicit any response (Rappole and Warner, 1980; Winker, 1987).

6.4.3 Mist-net results

We use mist-netting in addition to audio-visual sampling to assess habitat use by the wood thrush. For the 13 forested sites, the mist-net results from Costa Rica record a capture average of 3.0 wood thrushes/250 net-hours. Wood thrushes were not captured on five second-growth and five open sites (Tables 6.4, 6.5), and wood thrushes were not recorded in the VCP sampling in these same habitats (Table 6.3). Wood thrushes are not often found outside the forest in this region of Costa Rica.

Wood thrushes are captured more in second-growth habitat than in forest in Belize (Tables 6.4, 6.5). However, in the VCP sampling results, more are detected in forests than in second-growth habitats (Table 6.3). Results from both mist-net and VCP sampling for open sites show infrequent use of this habitat by the wood thrush in southern Belize. Finally, in Mexican forest sites, wood thrushes are found in intermediate abundances in relation to the Costa Rican and Belize sites (Table 6.4).

6.4.4 Examination of the habitat data

The winter habitat preferences for this species become clear when the remote sensing and census data are examined together with the available published information on distribution and habitat use. Wet forest is a preferable habitat for this species and the total amount of wet forest in the Costa Rica subscene is 1004 km^2. However, wood thrushes do not inhabit forest above 1000 m elevation in Costa Rica (Blake and Loiselle, 1992) and 45% (452 km^2) of the forests in this subscene are in this category. In addition, no wood thrushes were captured in three of the forested sites in Costa Rica (Table 6.4). All three of these sites are located in swamp forest, and virtually all forests located below the 50 m elevation contour in Costa Rica are swamp forests. In these forests, standing water covers the ground for a significant part of the year which makes this habitat of little use to wood thrushes that normally forage on the forest floor. In this

TABLE 6.4 The capture rates (number per 250 net-hour sampling period) and estimated minimum density of the wood thrush in the lowland evergreen forest (50–800 m elevation) study sites

Site number	Country		
	Mexico	Belize	Costa Rica
1	2	4	5
2	6	6	10
3	0	8	3
4	2	8	2
5	4	7	3
6	11	1	4
7	4	5	4
8	9	3	3
9	2	1	1
10	4	0	4
11	—	—	0
12	—	—	0
13	—	—	0
Total	44	43	39
Percent occurrence[a]	90	90	77
Mean/250 net-hours	4.4	4.3	3.0
Estimated minimum density/ha	2.4[b]	[2.3][c]	[1.6][d]

[a] Sample sites on which the species was captured/total sites sampled.

[b] Based upon radio-tracking of territorial individuals.

[c] Calculated using the formula $D_b = T_m/C_m (C_b)$, where D_b is the minimum density/ha of wood thrushes in Belize; T_m is density of wood thrushes in Mexico based on radio-tracking data; C_m is the mean number of captures/250 net-hours for the wood thrush in Mexico; and C_b is the mean number of captures/250 net-hours for the wood thrush in Belize.

[d] Calculated using the formula $D_c = T_m/C_m (C_c)$, where D_c is the minimum density/ha of wood thrushes in Costa Rica; T_m is the density of wood thrushes in Mexico based on radio-tracking data; C_m is the mean number of captures/250 net-hours for the wood thrush in Mexico; and C_c is the mean number of captures/250 net-hours for the wood thrush in Costa Rica.

study, swamp forest accounted for 24% (242 km²) of the total subscene forest.

The wood thrush in Belize is restricted to evergreen forest and second growth below 750 m (see Honduras data in Monroe, 1968). Additionally, in our samples the species occurs in much lower numbers (1.3 birds/ha) in stunted forests on karst slopes than in the wetter forests with better soils (5.7 birds/ha). Of the 10 randomly selected sites, three are in karst forest. If our sample reflects the relative abundance of karst forest in the subscene, then 30% of the available forest habitat is of low quality for wood thrushes.

Available habitat for wood thrushes is similarly restricted in the Tuxtla Mountain region of southern Mexico where our wood thrush surveys are located. While wood thrushes are found at densities of 2.4 birds/ha in our forested Mexican study sites, they are found very infrequently in open areas (Rappole et al., 1989, 1992). They are also absent from forests above 500 m in elevation (Winker, 1987). Forest cover in the San Martin portion of the Tuxtla region was 849.6 km² prior to European settlement (Dirzo and Garcia, 1992). In the 450 years since settlement, 713.5 km² of forest have been converted to open habitats,

TABLE 6.5 The capture rates and estimated minimum density of wood thrushes in lowland second growth and pasture (50–800 m elevation) at two Middle American study areas. The numbers represent the number of wood thrushes captured at each site during the 250 net-hour sample period

| | Country | | | |
| | Belize | | Costa Rica | |
Site number	Second growth	Pasture	Second growth	Pasture
1	2	1	0	0
2	0	1	0	0
3	17	2	0	0
4	7	0	0	0
5	8	0	0	0
6	9	1	—[a]	—
7	3	0	—	—
8	3	1	—	—
9	1	0	—	—
10	4	0	—	—
Total	54	6	0	0
Percent occurrence[b]	90	50	0	0
Mean/250 net-hours	5.4	0.6	0	0
Estimated minimum density/ha	[2.9][c]	[0.3]	0	0

[a] Only five second-growth and pasture sites were sampled in Costa Rica.
[b] Sample sites with individuals present/total sample sites.
[c] See Table 6.4 for formula used to calculate minimum density.

6.5 Modeling and analysis

mostly (50%) within the past 30 years (Dirzo and Garcia, 1992). Of the remaining 136.1 km² of forested habitat, only 20.6% (28.04 km²) is located below the 500 m contour.

Mist-net grid data do not provide a direct estimate for bird density. However grid data, in combination with information about the number of wood thrush territories/ha derived from intensive studies performed in many of the same areas in which the grid data were collected, may provide a useful index for the density of this species.

Many wood thrushes defend individual territories (Rappole and Warner, 1980; Winker et al., 1990a, b; Rappole et al., 1992) from the time of their arrival in tropical forest habitats in October until their departure in April. These territories have an average density of 2.4/ha in the lowland rainforests of southern Veracruz as determined through home range analysis of radio-tracking data (Winker, 1987; Winker et al., 1990a). We use this figure to calculate an estimated density of territorial wood thrushes/ha in Belize and Costa Rica (Tables 6.4, 6.5). We assume that a direct relationship exists between the number of captures on a 1-ha grid (250 net-hours) and the density of territorial individuals.

The difference between the number of territorial wood thrushes documented on a 1-ha site (generally 1 or 2 birds) and the total captures recorded in 250 net-hours on a 1-ha site (3–6 birds) may be due to the presence of non-territorial birds in search of available territories (Rappole et al., 1989). The number of non-territorial birds decline through the course of the winter because: (1) they locate vacant

TABLE 6.6 Estimated former and current wood thrush population sizes on study areas in Costa Rica, Belize and Mexico

Country/habitat	Est. former cover (km²)	Est. current cover (km²)	Est. density[a]	Former pop. size (×1000)	Current pop. size (×1000)
Costa Rica[b]					
Forest[c]	1296	310	210[d]	272.2	65.1
Second growth	13	578	0	0	0
Open	13	425	0	0	0
Other	—	9	—	—	—
Total	1322	1322		272.2	65.1
Belize[e]					
Mesic forest[f]	556	416	303	168.5	126.0
Karst forest[g]	238	178	72	17.1	12.8
Second growth	8	79	290	2.3	22.9
Open	8	77	30	0.2	2.3
Other		60	—	—	—
Total	810	810		188.1	164.0
Mexico[h]					
Forest[i]	538.0	28.0	240	129	6.7
Open	0.4	510.4	0	0	0
Total	538.4	538.4		129	6.7

[a] Calculated using formula explained in Table 6.4.
[b] Habitats are based on total area <1000 m in elevation.
[c] Forest habitats with canopies >10 m tall located between 1000 and 50 m elevation.
[d] Swamp forest sites are excluded. Number of birds/km²/250 net-hours = 3.9.
[e] Habitats below an elevation of 750 m.
[f] Seven mesic forest sites were sampled. Number of birds/km²/250 net-hours = 5.6.
[g] Three karst forest sites were sampled. Number of birds/km²/250 net-hours = 1.3.
[h] Based on Landsat habitat data for the San Martin portion of the Tuxtla Mountains, Mexico from Dirzo and Garcia (1992).
[i] Defined to include second-growth habitats >3 m tall. See text for explanation.

territories in preferred habitat (lowland tropical forest); (2) they settle in marginal habitat (lowland second-growth forest – 3–10 m canopy); or (3) they die of starvation and predation.

The remote-sensing data that document habitat type, abundance and distribution from the three countries are examined in relation to the estimated number of wood thrush territories/ha. This procedure provides an estimate of the total population sizes within each area (Table 6.6). These estimates are approximate, and they are based upon some density and habitat assumptions. Nevertheless, they provide a first-order approximation of the conservation threat confronting this species throughout its winter range.

Available winter habitat for this species has been significantly reduced in each of the three sites. Southern Belize includes the best remaining conditions of the three areas examined. The estimated decline in population of wood thrushes in southern Belize is only 13% since the pre-Columbian era. Contemporaneously, winter wood thrush populations have declined by an estimated 77% in northeastern Costa Rica, and by 95% in the San Martin region of the Tuxtla Mountains, Mexico.

6.6 Conservation applications

The estimates of wood thrush population declines at the three widely separated sites within its neotropical winter range highlight a serious threat to the survival of this species. Preferred habitats include low to mid-elevation wet forest and distinct second-growth patterns. These are the same areas that are most seriously impacted by human forest conversion.

The progress of land alteration is clearly reflected in the three sampled areas. In Belize, with a relatively low human population (8.2 individuals/km^2), there are still parts of the country where considerable amounts of forest and second growth remain. Costa Rica and Mexico have much higher human population densities, respectively 59.0 and 46.4 individuals/km^2, and correspondingly less lowland forest areas remain in these countries (World Resources Institute, 1992: 262). In southern Veracruz, an area with very high human population levels, almost all of the forest below 500 m is now cleared (95%). This results in the near extirpation of the wood thrush from this region.

The wood thrush is only one of thousands of species of animals and plants in the lowland evergreen broadleaf forest community of Middle America that is at risk due to forest conversion. The levels of forest loss seen at our Costa Rican and Mexican study sites are typical of forest-loss rates elsewhere in Middle America (Sader and Joyce, 1988; World Resources Institute, 1992). Dirzo and Garcia (1992) estimated the current rate of annual deforestation in the Tuxtlas to be 4.3%. Wet lowland forests in Mexico are reduced to an estimated 111 000 km^2 (Garcia and Perez, 1991) and current estimates of continued loss range from 4600 km^2/year (Repetto, 1988) to 16 000 km^2/year (Grainger, 1984).

6.7 Conclusions

Remote-sensing technology furnishes a powerful tool for assessing amounts and rates of change for winter habitats of migratory birds. When these data are used in conjunction with data from extensive ground surveys and data about home range size and survivorship, this approach can provide reliable measures of availability for a species' winter habitat.

This study does not resolve the issue of migratory bird 'plasticity' with regard to habitat use. However, the study does identify the preferred habitats of the wood thrush in its winter range. These preferred habitats are the lowland evergreen forest in Costa Rica, Belize and Mexico, and some forms of second-growth forest in Belize. Also identified are readily available but 'avoided' habitats where the wood thrush is found infrequently. These habitats include lowland second growth in Costa Rica (with 3–10 m canopies) and open areas with canopies less than 3 m in all three countries.

The remotely sensed imagery and mapping approaches used in this study permit us to identify the vulnerability of the wood thrush based on range limitations. The use of habitat by wood thrushes is inversely related to habitat availability. Therefore we conclude that the wood thrush lacks sufficient 'plasticity' for it to survive continued extensive alterations to its preferred winter habitats.

6.8 Summary

The wood thrush *Hylocichla mustelina* winters in the Caribbean lowlands of Middle America, an area that has undergone major deforestation in recent decades. We use Landsat imagery to identify vegetation types in northeastern Costa Rica and southern Belize, combining the different types into three classes after analysis: (1) forest, (2) second growth and (3) open. Randomly generated points are visited in each country, photographed and then analyzed for habitat characteristics. From this sample we select sites that represent each of the major vegetation classes. These are the sites on which we establish 1-ha mist-net grids and perform audio-visual censuses. We also include previously gathered data on wood thrush distribution and habitat use from southern Veracruz, Mexico. Wood thrushes occur primarily in forest, though birds are also found in second growth in Mexico and Belize. Few individuals are detected in habitats below 50 m (swamp forest) in Costa Rica, on forested karst slopes in Belize, or on deforested sites. The species is found in its winter range at increasingly lower elevations with increasing latitudes: <1000 m elevation in Costa Rica, <750 m in Honduras and <500 m in Mexico. Remote sensing is a powerful tool, and when it is combined with extensive ground censuses, it can provide estimates of habitat distribution and rates of change that are extremely useful for determining the conservation status of the wood thrush.

Acknowledgements

We thank Mr Robert Belisle for his assistance with Belize wildlife permits, and Major Andrew Duncan for technical advice concerning conditions in Belize. Mr Guillermo Canessa provided assistance with wildlife permits in Costa Rica. We are grateful to Courtney Conway, Patricio Choc, Walter Van Sickle, Rafael Flores, Dave Swanson, Harriet Powell, Bill McShea and Brian Miller for their field assistance. Joe Spruce assisted with analysis of the remote-sensing data for Belize. The National Fish and Wildlife Foundation, World Wildlife Fund for Nature and the US Fish and Wildlife Service Patuxent Migratory Bird Office provided funding for the project. Sam Beasom, Director of the Caesar Kleberg Wildlife Research Institute, Yolanda de los Santos, and Becky Davis, Administrative Officer, provided help to JHR with administrative procedures at Texas A&I University.

References

Blake, J.G. and Loiselle, B.A. (1992). Habitat distribution patterns of neotropical migrants at La Selva Biological Station and Braulio Carrillo National Park, Costa Rica, in *Ecology and Conservation of Neotropical Migrant Landbirds* (ed. J.M. Hagan III and D.W. Johnston), Smithsonian Inst. Press, Washington, D.C., pp. 257–72.

DeGraaf, R. and Rappole, J.H. (in press) *Neotropical Migratory Birds*. Cornell University Press, Ithaca, New York.

Dirzo, R. and Garcia, M.C. (1992) Rates of deforestation in Los Tuxtlas, a Neotropical area in southeast Mexico. *Conservation Biology*, 6, 84–90.

Garcia, M.C. and Perez, G. (1991) [Environmental deterioration of terrestrial biotic resources.] *Map of the Republic of Mexico*, Vol. 2.8, *National Atlas of Mexico*. Inst. Geografico, UNAM, Mexico, D.F, p. 8.

Grainger, A. (1984) Quantifying changes in forest cover in the humid tropics: overcoming current limitations. *Journal of Forest Resource Management*, 1, 3–63.

Green, K.M., Lynch, J.F., Sircar, J. and Greenberg, L.S.Z. (1987) Landsat remote sensing to assess habitat for migratory birds in the Yucatan Peninsula, Mexico. *Vida Silvestre Neotropical*, 1, 27–38.

Hutto, R.L. (1980) Winter habitat distribution of migratory landbirds in western Mexico with special reference to small, foliage-gleaning insectivores, in *Migrant Birds in the Neotropics* (ed. A. Keast and E.S. Morton), Smithsonian Inst. Press, Washington, D.C., pp. 181–204.

Karr, J.R. (1976) On the relative abundance of migrants from the north temperate zone in tropical habitats. *Wilson Bulletin*, 88, 433–58.

Lynch, J.F. (1989) Distribution of overwintering Nearctic migrants in the Yucatan Peninsula, I: General patterns of occurrence. *Condor*, 91, 515–44.

Monroe, B.L., Jr (1968). A distributional survey of the birds of Honduras. *Ornithological Monographs*, 7, 1–458.

Morse, D. (1971) The insectivorous bird as an adaptive strategy. *Annual Reviews in Ecology and Systematics*, 2, 177–200.

Petit, D.R., Petit, L.J. and Smith, K.G. (1992) Habitat associations of migratory birds overwintering in Belize, in *Ecology and Conservation of Neotropical Migrant Landbirds* (ed. J.M. Hagan III and D.W. Johnston), Smithsonian Inst. Press, Washington, D.C., pp. 247–56.

Phillips, A.R. (1991) *The Known Birds of North and Middle America*, Part 2, Allan R. Phillips, Denver, Colorado.

Rappole, J.H. and Warner, D.W. (1980) Ecological aspects of avian migrant behavior in Veracruz, Mexico, in *Migrant Birds in the Neotropics* (ed. A. Keast and E.S. Morton), Smithsonian Inst. Press, Washington, D.C., pp. 353–93.

Rappole, J.H., Morton, E.S. Lovejoy III, T.E. and

Ruos, J.R. (1983) *Nearctic Avian Migrants in the Neotropics*, US Fish and Wildlife Service, Washington, D.C.

Rappole, J.H., Morton, E.S. and Ramos, M.A. (1992) Density, philopatry and population estimates for songbird migrants wintering in Veracruz, in *Ecology and Conservation of Neotropical Migrant Landbirds* (ed. J.M. Hagan III and D.W. Johnston), Smithsonian Inst. Press, Washington, D.C., pp. 337–44

Rappole, J.H., Ramos, M.A. and Winker, K. (1989) Wintering Wood Thrush movements and mortality in southern Veracruz. *Auk*, **106**, 402–10.

Repetto, R. (1988) *The Forest for the Trees? Government Policies and the Misuse of Forest Resources*, World Resources Inst., New York.

Reynolds, R.T., Scott, J.M. and Nussbaum, R.A. (1980) A variable circular-plot method for estimating bird numbers. *Condor*, **82**, 309–13.

Robbins, C.S., Sauer, J.R., Greenberg, R.S. and Droege, S. (1989) Population declines in North American birds that migrate to the Neotropics. *Proceedings of the National Academy of Science*, **86**, 7658–62.

Sader, S.A. and Joyce, A.T. (1988) Deforestation rates and trends in Costa Rica, 1940 to 1983. *Biotropica*, **20**, 11–19.

Sader, S.A., Powell, G.V.N. and Rappole, J.H. (1991) Migratory bird habitat monitoring through remote sensing. *International Journal of Remote Sensing*, **12** 363–72.

Tosi, J. (1969) *Mapa ecologico de Costa Rica*, Tropical Science Center, San Jose, Costa Rica.

Winker, K. (1987) The Wood Thrush (*Catharus mustelinus*) on its wintering grounds in southern Veracruz, Mexico. M.S. thesis, University of Minnesota, Minneapolis, Minnesota.

Winker, K., Rappole, J.H. and Ramos O., M.A. (1990a) Population dynamics of the Wood Thrush in southern Veracruz, Mexico. *Condor*, **92**, 444–60.

Winker, K., Rappole, J.H. and Ramos O., M.A. (1990b) Within-forest preferences of Wood Thrushes wintering in the rainforest of southern Veracruz. *Wilson Bulletin*, **102**, 715–20.

World Resources Institute (1992) *World Resources: 1992–93*, Oxford University Press, New York.

Part Five

Using Maps for the Conservation of Large Mammals Around the Globe

An increasing number of international large mammal studies are using species distribution maps to convey their message about conservation priorities (e.g. Hatough-Bouran and Disi, 1991; Smith and Mishra, 1992) and elephant conservation is a featured concern of conservationists world wide (e.g. African Elephant Conservation Coordinating Group, 1991; DFBC, 1991; FDA 1991; Lindeque, 1991; Campbell and Huish, 1992; Environmental Investigation Agency, 1992; National Parks Board of South Africa, 1992; Phanthavong and Santiapillai, 1992). The decade leading to the international ban on the ivory trade in 1989 witnessed a halving in the wild population of the African elephant, *Loxodonta africana*. Poaching for ivory, the destruction of suitable habitat, and human population expansion are all factors that have contributed to the decline of this species. The African Elephant Database (AED) introduced in Chapter 7 arose in response to a growing need by conservationists and development agencies alike for sound and reliable elephant data for both development and conservation planning.

Chapter 7 presents a clear assessment of the key data quality problems encountered during the building of the AED. These same issues are being confronted repeatedly when data are taken from scientific disciplines and used to address real world applications. Some up-to-date examples of practical field uses for the Global Positioning System (GPS) technology in Africa are also discussed in this chapter. More widespread use of this technology in field work in Africa will rapidly improve the geographic data quality of both species and habitat data being collected in Africa.

For the past six years, the AED has served as the primary foundation for a scientific consensus on the status of the African elephant. This database, regularly updated as new data are received, gives timely and consolidated assessments of elephant populations for the biennial Convention on International Trade in Endangered Species (CITES). Recent database developments are enabling users to deal with a broader range of wildlife planning and management issues at the regional, national and local levels. In addition, the AED is becoming an important tool for the identification of conservation priorities at the tactical, strategic and policy levels (Norton-Griffiths *et al.*, 1991). In Chapter 7 two case studies are presented that illustrate the broader potential of the AED.

Chapter 8 introduces a modern approach that uses statellite imagery and GIS for designing effective protected areas for the giant panda in China. As a result of human encroachment, the range of the giant panda is now condensed to just six isolated areas. In addition, over the past 20 years, the panda's chief food resource — arrow bamboo — has also declined steadily. Even though many reserves have been established by the Chinese government, panda numbers decline steadily. Over the last 20 years two major field surveys were done together with detailed mapping of forest cover derived from satellite imagery data. The results of these surveys show that the suitable habitat occupied by giant pandas is continuously declining. Recent use of remotely sensed time sequence mapping, described in Chapter 8, has proven invaluable for identifying the problems and possible solutions needed to save the giant panda.

Between 1986 and 1989 WWF experts worked closely with officials of China's Ministry of Forestry to prepare a master plan for saving the giant panda and its habitat. Use of satellite imagery from the habitat of the giant panda proved to be of vital importance for understanding the crux of the real problems affecting this species. These imagery data also served to focus the viewpoints of the Chinese and foreign scientists involved in the planning and in the identification of the management solutions.

References

African Elephant Conservation Coordinating Group (1991) *The African Elephant Conservation Review*, AECCG, 21 St Giles, Oxford, UK, 66pp.

Campbell, K. and Huish, S. (1992) *Recent Trends in Tanzanian Elephant Populations: 1987–1992*, TWCM, Arusha, Tanzania, 7pp.

DF & C (1991) *Elephant Conservation Plan: Gabon*, Direction de la Faune et de la Chasse, Libreville, Gabon, 77pp.

Environmental Investigation Agency (1992) *Under Fire: Elephants on the Front Line*, Environmental Investigation Agency, London, 57pp.

FDA (1991) *Elephant Conservation Plan: Liberia*, Forestry Development Authority, Monrovia, Liberia, 47pp.

Hatough-Bouran, A. and Disi, A.M. (1991) History, distribution, and conservation of large mammals and their habitats in Jordan. *Environmental Conservation*, 18 (1), 19–32.

Lindeque, M. (1991) *Elephant conservation and management plan: Namibia*.

National Parks Board of South Africa (1992) *Draft Elephant Conservation Plan for South Africa*, 43pp.

Norton-Griffiths, M., Campbell, K. and Michelmore, F. (1991) Applications for Geographic Information Systems in Wildlife Management, in *Wildlife Research for Sustainable Development*. Proceedings of an International Conference held in Nairobi, Kenya, by the Kenya Agricultural Research Institute (KARI), the Kenya Wildlife Service (KWS) and the National Museums of Kenya (MMK), April 22–26, 1990.

Phanthavong, B. and Santiapillai, C. (1992) *Conservation of Elephants in Laos*, IUCN/SSC Asian Elephant Specialist Group Newsletter 8, pp. 25–33.

Smith, J.L.D. and Mishra, H.R. (1992) Status and distribution of Asian elephants in central Nepal. *Oryx* 26 (1; January), 34–8.

Chapter seven

Keeping elephants on the map: Case studies of the application of GIS for conservation

Frances Michelmore

7.1 Introduction to this approach

The African Elephant Database (AED) is an important species-monitoring program within the Global Environment Monitoring System (GEMS) of the United Nations Environment Programme (UNEP). It is based within the Global Resource Information Database (GRID) office of the UNEP headquarters in Nairobi, Kenya. The AED enables these programs to synthesize and present a continental assessment of the status and distribution of this species from the available information (Figure 7.1). This database is perhaps the most comprehensive and geographically extensive GIS concerned with a single species. However, the widespread distribution of the elephant requires the compilation and assessment of information that varies widely in its quality and availability. A more robust basis for conservation planning to ensure the ultimate survival of the African elephant will result when we master the problems related to mapping and data accuracy.

Problems involving conflict between people and wildlife are more numerous and complex due to increasing human populations and the concomitant demands for land to produce food and build livelihoods. These difficulties are prevalent across the globe in spite of the significant attempts made over the past two decades to bring together the needs of conservation and the goals of development. Critical

Mapping the Diversity of Nature. Edited by Ronald I. Miller.
Published in 1994 by Chapman & Hall, London. ISBN 0 412 45510 2.

FIGURE 7.1 Range of the African Elephant, *Loxodonta africana*, across the African continent (after Douglas-Hamilton *et al.*, 1992).

areas of habitat are rapidly being fragmented and eroded by competing land uses. Vast quantities of disparate spatial data that document these changes are currently available. The use of the GIS is the practical and efficient avenue today to systematize, standardize and manage the enormous amounts of spatial data generated by disturbance to the landscape.

The gap between the practical side of data acquisition and the processing powers of computers is being closed by the availability of automated mapping technologies and the improved accuracy and precision of the data collection methods. Practical skills and experience used together with advanced mapping and navigation tools promote the successful and sustainable implementation of effective conservation programs.

The most difficult tasks linked to the use of the GIS as an analytical tool involve the identification of the problems, the definition of the objectives, and the selection of the appropriate GIS methods for the task. To answer these fundamental questions we need to decide what the final use of the data will be, and who will be using them. For example, if the elephant data are being collected to detect population abundance trends, this will require less of an effort than that required to determine either the impact of resource use or the impact of human activities on elephant populations at national or local levels. Data can easily be misinterpreted if their resolution and quality are not considered. This is a particular danger with GIS, where it is easy to be convinced by the seductive maps that appear on the computer screen. There is a tendency among users to expect digital data to be of higher quality than conventional map data or data held in manual files. This is not the case and it is necessary to keep in mind that the analysis and interpretation of the data depend largely upon the users' capabilities and not the GIS technology.

7.1.1 The continental assessment

The AED continues to thrive based upon the solid foundation established by Iain Douglas-Hamilton and others in 1986 and the continuing technical support provided by the World Conservation Union's (IUCN) African Elephant Specialist Group (AESG). This group has met periodically since 1975 to review available technical information derived from surveys, questionnaires, reports and anecdotal sources (Douglas-Hamilton, 1979; Cumming and Jackson, 1984; Burrill and Douglas-Hamilton, 1987).

In order to develop plans for their conservation and management, it is necessary to understand how elephants are distributed. Elephants are large and conspicuous, and thus data on their distribution can be gathered with relative ease from both ground and aerial surveys. However, the fact that they may travel great distances and are found in most habitat types across Africa can also present problems. Data collection over large areas or within thick bush and forest is often difficult.

GIS tools permit maps to be overlaid at the same scale and in a common coordinate system. Those factors most closely correlated with elephant numbers and distribution can be identified and used to predict elephant numbers in areas without a history of previous survey work. This is accomplished by using a GIS for integration and analysis of environmental elements associated with the elephant (Burrill and Douglas-Hamilton, 1987; Michelmore *et al.*, 1992).

In addition to holding data on the present (1992) range and numbers of the African elephant, the database stores information on the network of protected areas in the Afrotropical region and on predominant vegetation types. Through the World Conservation Monitoring Centre (WCMC) in Cambridge, UK, the African elephant database is able to acquire up-to-date information relating to the status and extent of gazetted national parks, game reserves, and other protected areas in the Afrotropical region. This data layer can then be overlaid with elephant range. Given the critical importance of protected areas for the survival of elephant populations throughout Africa (Burrill and Douglas-Hamilton, 1987), this information is invaluable for assessing the adequacy of protection afforded to the African elephant and for identifying the different vegetation types falling within its range.

Recent analyses indicate that in many parts of Africa the protected area system is not conserving the spectrum of biodiversity (Douglas-Hamilton *et al.*, 1992). A mere 1.5% of the total range of the African elephant is adequately protected and approximately 76% of the range of this species is not protected at all. Of the forests of central Africa – covering some two million square kilometers in six countries – only 1% are afforded any protection at all. This is inadequate considering the unique and rich biodiversity associated with these forests. African reserves and parks were traditionally created to protect either the charismatic high-profile species or the large concentrations of game (Mackinnon and Mackinnon, 1987). Consequently, what we have today are many areas of exceptional ecological richness that lack any kind of protection.

The AED is valuable for identifying patterns of change at continental and regional levels. It accomplishes this by using the overlay capabilities of the GIS to contrast stored data layers. For example, the fragmentation of habitat and the increasing isolation of elephant populations in West Africa is now also becoming evident in East Africa. These striking changes are evident within the lifetime of the AED and they certainly are noticeable within the two-decade period that good elephant data have been collected in Africa (I. Douglas-Hamilton, pers. comm., 1992).

Limitations in the elephant data make comprehensive trend analysis impossible. However, where key elephant populations are regularly censused over a period of a decade or more, these data may produce a sample trend indicative of the broader picture for a region.

7.1.2 The acquisition, storage, manipulation and output of data

The AED uses a number of computer-based systems for the compilation, analysis and output of elephant information. The Environmental Systems Research Institute's (ESRI) ARC/ INFO software is the main GIS system used to store, analyze and display the information on both mainframe and PC computer systems at the GRID–Nairobi facility. For the purposes of providing a continental overview, national maps commonly at a scale of 1:1 million are computerized and transformed for incorpora-

tion into a continental map (Michelmore, 1991).

Spatial information within the AED is stored in vector format in which features are represented by points, lines and polygons. Raster-based source data are stored as uniform, systematically organized cells (e.g. a grid system) and these are transformed and incorporated into vector information (Lamprey *et al.*, 1991).

The source data for the AED include maps of differing scales, map projections, quality and accuracy. Data are also received in tabular format or in the form of questionnaire replies. Other sources include data extracted from reports and articles. All these source data types are integrated by the GIS. Standardized storage of the data using common reference systems and uniform formats simplifies the presentation of the data from these dissimilar sources. For example, versions of the database at both continental and national scales can produce time-series analyses which illustrate changes in elephant ranges and densities. The analyses results may sometimes represent real population changes and sometimes they represent improvements in census techniques which produce more accurate survey results. Consequently, results from GIS analyses should always be interpreted with caution.

7.1.3 Data quality problems

Many problems are associated with the dissimilarity of spatial data. For example, the numerous georeference systems which describe the globe in different ways and with varying precision create problems in data interpretation. Adoption of a common map projection is necessary to compare data originating from maps at different scales. Similarly, transformation of map data to a common coordinate system is required if the data are to be used analytically. For instance, data received from field surveys using the Global Positioning System (GPS) must be referenced to a common coordinate system for overlay onto GIS base maps. GISs are generally equipped to store and process most varieties of geographically referenced data.

During the establishment of the AED, methods were sought to produce high-quality gray-scale maps for both report production and general distribution. This was particularly necessary since the expense involved to produce color maps would have seriously limited the capacity for map distribution. The technique selected interfaces the primary GIS (ARC/INFO) to an Apple Macintosh computer. Maps are converted and transferred to a graphics package on the PC without losing map detail or configuration. The maps are then ready for editing and final report-quality production.

The dissemination of updated information to the countries that contribute data on elephant distribution or abundance is an essential component of information management for the AED. The most effective format for this information is tabular estimates referenced to maps. This provides the wildlife management authorities in the range countries with the opportunity to approve the current contents of the elephant database. This also ensures that expert knowledge is used at all stages to control the validity of the data.

(a) Field data

The methods used to gather field data diverge between aerial and ground techniques of reasonable accuracy to methods that are inexact and conjectural. The quality of the data held in the database and the ability to accurately represent this wide variability is of prime concern to users and data managers alike. Elephant census is far from being precise. However, these are the only extant data from which the status, distribution and trends of the elephant can be determined. Therefore this is the principal resource for the formulation of policies for the future conservation and management of the African elephant.

Acquired data have to-date been assigned a rather crude and arbitrary quality rating on a scale from 1 to 3. These ratings include aerial surveys in category 1 (the highest grade), extrapolated data from mathematical modeling analyses in category 2, and informed guesses in category 3. There is a recognized need to improve the way that the quality of the original data is reflected in the output from the AED. If current improvements in this regard are successful, we will soon have a greater understanding of the differences in the methodologies used to collect data at regional and national levels. This will certainly improve the quality of the data stored in the database.

The effectiveness of the database depends on both the accuracy and precision of the data. Even in the case of the most sophisticated census techniques, there usually will be a bias which affects the accuracy of the collected data. On the other hand, precise data have a small (sampling) error and are therefore more readily compared with subsequent data.

Two phenomena that affect the precision and accuracy of survey estimates are errors and biases. Errors are random by definition, and therefore their cumulative effects cancel out. Errors are thus statistical and they are quantified by the confidence limits of a survey estimate. On the other hand, biases are methodological and produce error in a consistent direction. The effect of this is cumulative and increases the bias rather than canceling it out. In surveys, such biases must be controlled to maintain accuracy and precision. Biases can arise in the aerial survey design, from inaccurate maps, from aircraft piloting, from observer performance and from the analysis of the vertical aerial photography.

Aerial census data usually contain significant variance that cause inexact survey results. No matter how rigorously one controls the sources of bias, undetected sources will always exist that cannot be quantified or corrected. Most importantly in aerial surveys, bias from observational difficulties tend to result in underestimates of populations. The application of correction factors to reduce bias is imprecise because a lot of data are required to identify these factors and the corrections often result in overestimates. Techniques are needed to improve census precision in aerial surveys and thereby to reduce variance.

(b) Map data

Apart from a small proportion of data collected using the Global Positioning System (GPS), most of the source data for the elephant database have originated from paper maps of varying quality. A number of cumulative errors are therefore present prior to digitization that arise from the original survey techniques, the photogrammetric interpretation, and the cartography of the pre-existing maps.

In traditional cartographic mapping, accuracy is inversely proportional to the scale of the map. However, the relationship is more complex when dealing with digital map data. Digital maps tend to be stored at a single scale irrespective of the scale in which they were collected and digitized. Maps can then be produced at any scale specified by the user. However, the data can only be wisely used for applications at the scale at which it was originally input (e.g. data entered at a scale of 1:1 000 000 cannot be used for analyses at a 1:50 000 scale).

Inaccurate georeferencing can result from the selection of inappropriate interpoint intervals during the process of digitization. These intervals determine the density of points used to represent linear features or areas on the map. Lines on a map with considerable detail along their length and large directional changes will be more accurate when a smaller interpoint interval is adopted.

Other sources of error include printing errors and non-uniform distortion of paper maps. More stable mylar maps are usually not available for digitization. These types of errors can

give rise to serious problems, particularly when joining digital map sheets.

Information is extracted from maps produced by various agencies using different map series and often different symbols are used for the same themes. Together with the concomitant errors that are produced during the digitization process, alignment disparities are common. Map data derived from these sources should include accuracy documentation. GIS users will then be able to assess the consequences of the various sources of error involved.

7.2 Case Study 1: A GIS model for estimating elephant numbers in the forests of central Africa

7.2.1 Introduction

GIS offers powerful modeling tools for wildlife managers. A spatial model was developed to predict elephant distribution and numbers in countries with particularly poor data. In these countries, enormous logistical difficulties were involved in conducting field survey work.

As documented by Barnes (in Cobb, 1989; Barnes and Barnes, 1992), the forests of central Africa have remained until recently the biggest challenge in the assessment of elephant populations in Africa. The rainforests of the equatorial region of Africa comprise some one-third of the range of the African elephant. These rainforests are contiguous across the borders of six central African countries: Zaire, Gabon, Equatorial Guinea, Cameroun, Congo and the Central African Republic (CAR) (Figure 7.2). Many field biologists and scientists attempt to penetrate these forests in the quest for knowledge about their rich biological resources. We perhaps neglected this important elephant range, secure in the assumption that these forests probably harbored and protected hundreds of thousands of elephants. For

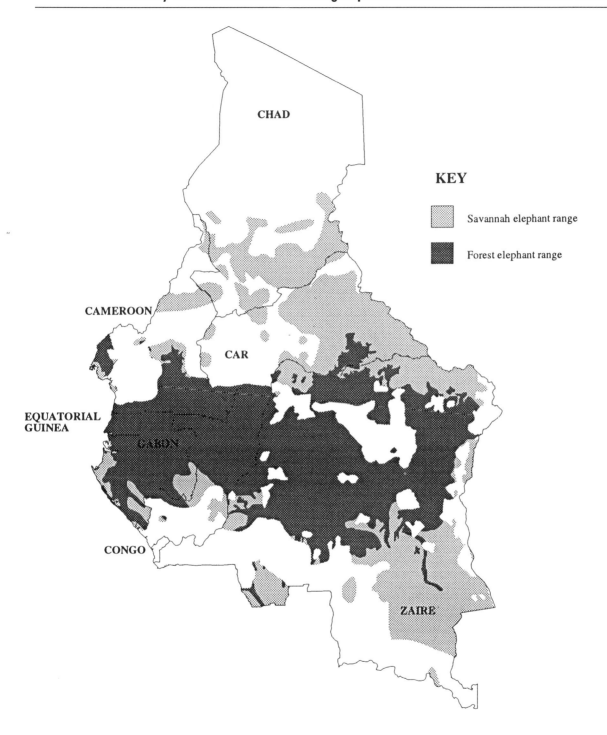

FIGURE 7.2 The distribution of the forest and savannah elephant range in central Africa (after Michelmore, 1991).

example, the forest elephant surveys organized by Wildlife Conservation International (WCI) have given a better idea of the true situation and have shown the impact of ivory poached.

7.2.2 Methodology

Wildlife surveyors have been using indirect methods since the early 1970s (Wing and Buss, 1970; Short, 1983; Merz, 1986; Barnes and Jensen, 1987) due to the logistical difficulties involved in trying to count elephants in forests. The cost of field survey work per unit area, relative to the quantity of data collected, is much greater in the forest than in savannah habitats and is also extremely time-consuming. Thus maximizing the usefulness of the information gathered in the forest is vitally important.

Estimating elephant numbers in forests where they are seldom seen involves determining elephant dung-pile density and establishing its relationship to elephant numbers. This method has enabled scientists to obtain realistic estimates of forest elephant populations for the first time (Jachmann, unpublished). Using multivariate analysis, the proximity of human disturbance was found to be the critical factor influencing the density and distribution of elephants within the forests of Gabon. Elephant densities were found to increase with distance from roads and major rivers where human settlements are concentrated.

Working with the AED, the field scientists were able to control the quality of the data entering the database. This simplified analysis and provided essential input during all phases of the GIS operation. We were thus able to extrapolate from a relatively small original sample of data to an area approaching two million square kilometers of forest elephant range. However, there are inherent dangers in extrapolating data to determine estimates over such large, unsurveyed areas. Therefore disclaimers should be attached to discourage

extrapolation when the data are inappropriate for such applications.

A 1:2 000 000 map series was chosen for the purposes of mapping roads and major rivers. Although these maps were originally published in the 1960s, they were thought to be more accurate than some of the larger-scale maps available (R.F.W. Barnes, pers. comm., 1989). The digital data layers (i.e. coverages) were created for each parameter and organized by theme. These data layers included elephant range, elephant density, roads, rivers and vegetation. The vegetation data were further divided into subsidiary layers for forest and savannah types. In addition, the roads had a unique identification code attached to them according to the grade or type of road (Michelmore et al., 1992).

Maintaining parameters in separate data layers facilitates the updating of the database. For example, elephant range, probably the most frequently updated layer, is stored separately. This storage of data in thematic layers, so that each data layer involves a single theme (e.g. soils, vegetation, land-use, etc.), is analogous to the overlays of conventional map making. For example, road types can be selectively removed from a hierarchical layer of the road map should they become upgraded or disused in time (e.g. in the case of logging trails).

7.2.3 Modeling and analysis

Gabon is presented here to illustrate aspects of the modeling and analysis procedures. The number of elephants in Gabon can be calculated from the number of dung-piles, their decay rate and the apparent rate of defecation. It is assumed that defecation and decay rates are constant and that the sample study area is in a state of equilibrium in terms of the number of elephants entering and leaving the system. Therefore the estimates will be inaccurate if the area is colonized or if decay or defecation rates fluctuate (e.g. as a result of variation in rainfall). These inaccuracies will be particularly

White bands indicate areas of few or no elephants in the vicinity of roads and major rivers

Large tracts of forest with high elephant estimates.

Contours at 4 km intervals through the forest, representing bands of increasing elephant density away from roads and major rivers.

Large white areas are savannah zones or areas of non-range for elephants.

FIGURE 7.3 A GIS map indicating the areas of forest elephant range in Gabon that contain potentially high biodiversity. The 4 km wide bands between the contours represent areas of increasing elephant density. These are forest areas away from roads and major rivers.

conspicuous if they are used to extrapolate over large areas (McClanahan, 1986; Barnes and Jensen, 1987; Barnes and Barnes, 1992). The AED was used to estimate elephant numbers for the whole country (Barnes *et al.*, 1992) once a mathematical relationship between dung-pile density and distance to the nearest road or major river was established in Gabon.

Buffer zones that defined the spatial proximity of forests to roads and major rivers (i.e. those navigable by humans) were generated by the GIS at 4 km intervals in forested areas. These zones were compiled automatically by the computer. The areas within 7.5 km on each side of roads and rivers were found to contain few elephants during the course of field surveys. These areas were assigned zero elephant density for the purposes of the computer-modeling exercise (Figure 7.3). The superimposed data layers of interest produced an estimate of the number of elephants that occurred between each zone. Summation of these estimates yielded an extrapolated total elephant population for the entire country.

There are many potential sources of error involved in arriving at a final estimate of elephant numbers using this method. The most important sources include the following:

- the variable rate of elephant defecation in forest: however this falls within comparatively narrow limits in contrast to the savannah environment and it exhibits little seasonal variation (Wing and Buss, 1970; Merz, 1986; Tchamba, 1992);
- dung decay rate: the rate depends upon a variety of factors, such as habitat type, rainfall, temperature and humidity;
- sample sizes for all countries, with the exception of Gabon, were small; consequently, large errors could have been introduced in determining elephant numbers by extrapolation over large areas;
- the correct identification of navigable rivers: for this model the identification was arbitrary and based on knowledge in limited areas;
- no distinction was made between dryland and swamp forests; however, recent findings (J.M. Fay, pers. comm. to R.F.W. Barnes) indicate that elephant numbers may well be lower in the swampy habitat;
- recent findings (Alers *et al.*, 1992) show that elephants no longer exist in large tracts of forest in Zaire primarily as a result of poaching for ivory; however, the model for Zaire did not take this into consideration since the extent of these areas is still unknown.

On the basis of the last two statements, current GIS estimates based on this model are likely to err on the optimistic side.

7.2.4 Conservation applications

The forest elephant GIS model represents the most important recent development in forest elephant survey techniques (R.F.W. Barnes, pers.

comm., 1992). This modeling approach has helped to clarify the status of elephants in the central African forests. In particular, the model serves to illustrate the magnitude of the regional decline in forest elephant numbers. The GIS tools enable us to draw comparisons between undisturbed and heavily poached populations. It suggests that the total elephant populaton of the equatorial forest region is reduced by more than 40%, primarily as a direct consequence of ivory poaching (this assumes that elephants originally ranged throughout the forest). Cameroun, Equatorial Guinea, Zaire and the Central African Republic are believed to have lost approximately half of their forest elephants, and Congo approximately one-third. The reduction in numbers is also due to range contraction brought about by the loss of suitable habitat and human disturbance (Michelmore *et al.*, 1989).

The maps illustrate the likely concentrations of elephant populations (Figure 7.3) and provide the concerned governments with a much better understanding of their elephant populations. In addition, the maps clearly indicate forested areas of potential importance for elephants. This helps to provide a conceptual framework for the efficient design of further field surveys. In addition, we now have an indication that elephants may be more vulnerable to poaching in certain areas than previously thought. For example, the GIS model enables us to determine that more than 60% of Congo's elephants live within two days' walk (i.e. 11–40 km) of a road or navigable river (Michelmore *et al.*, 1989).

One of the advantages of GIS is the ease with which the African elephant database can be updated with the acquisition of new data. It is hoped that better estimates can be obtained for each country that will improve the GIS model. In addition, there is a need to acquire data about other activities within the forest (e.g. logging intensity) so that the estimates of elephant numbers can be further improved.

7.3 Case Study 2: Monitoring large elephant herds in Kenya

7.3.1 The Laikipia Elephant Project: Resolving the human/elephant conflict for land

Conflicts between humans and wildlife for natural resources are but one symptom of the world's rapidly growing human population and the concomitant increase in demand for food and land. In the area of Kenya discussed here, poaching may have accelerated the process leading to direct conflict between humans and elephants. The increase in poaching in northern Samburu District forced elephants to seek refuge further south on the Laikipia Plateau (Thouless and Dyer, 1992). Elephants initially sought temporary refuge when they were driven south by the onslaught of poachers. Later the elephants became predominantly resident on large private ranches on the Laikipia Plateau, with seasonal movements away from this secure environment only at certain times of the year. Later the ranches were subdivided and sold as small-scale farms and the elephants have increasingly come into conflict with farmers through the destruction of crops and buildings.

The purpose of the KWS/WWF Laikipia Elephant Project* (LEP) is to determine and interpret the movements of elephants in order to find ways of reducing and resolving the conflict. Important digital thematic data layers (e.g. hydrology and land use) were used in this project.† The use of these data saved the LEP valuable time and manpower. Unnecessary duplication of effort is a potential problem which plagues the world of GIS. Different organizations or projects are often engaged in the digitization of maps for the same areas of interest.

The weekly movements of 17 radio-collared elephants, of an estimated total population of 2500 in the area, were superimposed on the thematic data layers of interest (e.g. land use, protected areas, hydrology, roads, elevation, land ownership, vegetation) for a period of two years. Conventional radio-telemetry techniques were used together with a Global Positioning System (GPS) to track the animals. Figure 7.4 illustrates the movements of one of these collared elephants between October 1990 and January 1992. The reliability of the radio-collars (produced by Telonics Inc.) together with the regular and successful tracking operations produced one of the most extensive studies of a migratory elephant population in Africa (Thouless and Dyer, 1992).

7.3.2 Increased precision in elephant census: Satellite navigation systems and GIS

The GPS technology uses dedicated orbiting satellites to determine the position of the user on the ground, sea, or in the air. Based upon the classic technique of triangulation, GPS uses this series of satellites as fixed points in space to perform the positioning calculations. The constraints experienced by surveyors on the ground do not exist in space as the satellite information is accessible regardless of weather conditions. The incorporation of such systems within survey aircraft is straightforward and is already being used with notable success. For example, GPSs were used experimentally for the first time in an aerial census of elephants in Tsavo National Park, Kenya (Figure 7.5). In addition, it is now possible to map simultaneously the results of GPS surveys onto preloaded GIS base-maps of the area being surveyed (Figure 7.6).

7.3.3 Conservation applications

Combining the techniques of GIS and GPS, these projects have produced some interesting

* The LEP is coordinated by Dr Chris Thouless, directed by the Kenya Wildlife Service (KWS) and funded by the World Wide Fund for Nature and Bunzl Plc, in collaboration with the Zoological Society of London and the Gallmann Memorial Foundation.
† These data were received from the Swiss funded Laikipia Research Project, based in Nanyuki, Kenya.

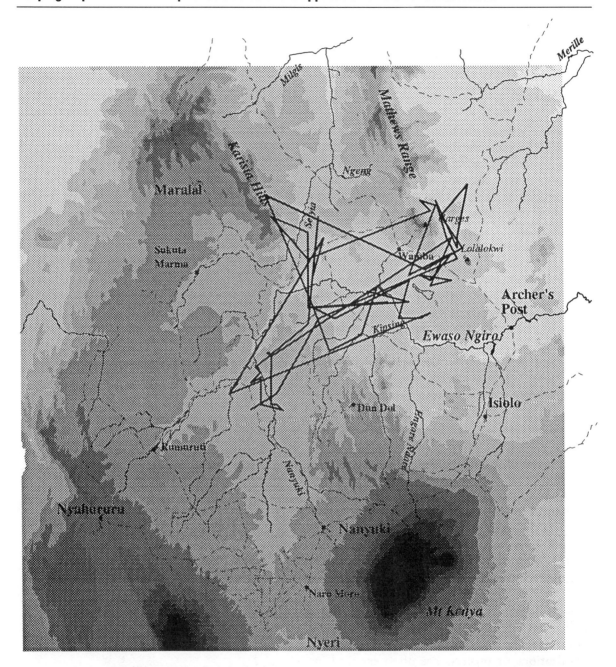

FIGURE 7.4 A GIS map of elevation in Laikipia District, Kenya. The movements of one radio-collared elephant are superimposed upon this area.

and potentially exciting results. These results pave the way for the application of these tools to elephant survey and management problems.

There is a great deal that can still be done in order to minimize bias in elephant surveys. One day GPS may still revolutionize the way

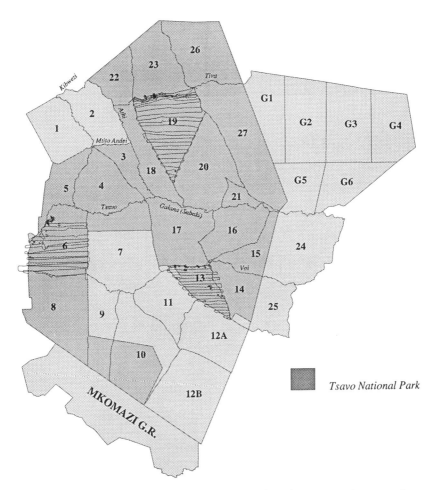

FIGURE 7.5 A GIS map of ecosystems located in Tsavo National Park, Kenya. This map illustrates the block divisions for aerial census of elephants. The flight paths of aircraft recorded with a Trimble Pathfinder GPS are presented in three experimental blocks.

that we carry out aerial surveys and also perhaps ground surveys within forests. The use of this technology will help to reduce a number of the aforementioned biases. GPS virtually eliminates bias from gross errors in navigation. In addition, it allows the pilot to monitor ground speed very accurately and to adjust the air speed of the aircraft accordingly. For example, aircraft tend to travel faster across the ground when flying downwind as opposed to upwind. Changes in aircraft speed can represent an important source of error insofar as observer counting bias increases with the ground speed of the aircraft (Clarke, 1986).

The link between GPS and a field-based GIS system (e.g. **Geolink**) allows data to be collected in digital form from the field in a preselected, common reference system. The data are thus ready to transfer with ease into other GIS systems.

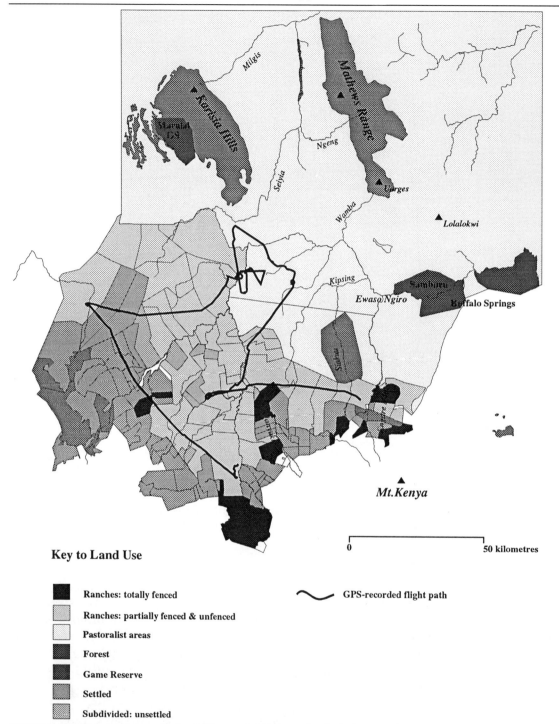

Key to Land Use

⬛	Ranches: totally fenced
▨	Ranches: partially fenced & unfenced
☐	Pastoralist areas
▨	Forest
⬛	Game Reserve
▨	Settled
▨	Subdivided: unsettled

〜 GPS-recorded flight path

FIGURE 7.6 This map presents the GPS recorded flight path of a light aircraft with radio-tracking equipment. The locations of radio-collared elephants were mapped simultaneously onto GIS-generated basemaps of Laikipia Plateau, Kenya (after Thouless and Dyer, 1992).

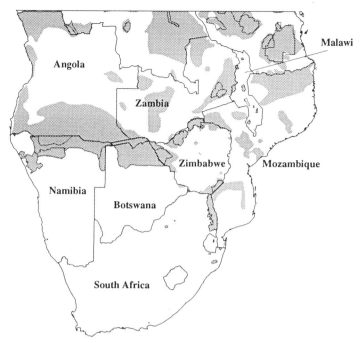

FIGURE 7.7 This map shows all protected areas overlaid upon elephant ranges in southern Africa (after Douglas-Hamilton *et al.*, 1992).

7.4 Conclusions

The AED is the only reference source that gives readily available numbers relevant to the known distribution of the African elephant. This enables the scientific community to make a continental assessment that is particularly important for the policy decision makers. However, continental initiatives to tackle regional or national projects are inappropriate; most resource management projects are implemented at the national level. Refinement of the database to meet national requirements will strengthen the value and reliability of the continental picture. In response to the needs of the user community, the custodians of the AED are now storing data to enable elephant conservation and management issues to be addressed at a finer level of detail.

The AED offers a good demonstration of the implications of presenting the same data at different scales. For example, within Kenya the details of land use in the Laikipia district can be used for the resolution of elephant/human conflicts and therefore for conservation planning and elephant management at a population level (Figure 7.6). At the national level, however, the fine details of these same distribution data are lost and the resulting information is used more for strategic and national policy planning. At the continental scale, all such detail is lost. However, regional problems (e.g. the management of cross-border elephant populations) subsequently become apparent.

In their quest for safe refuges to feed and water, elephants do not recognize international borders. As a result, a number of countries may share a particular population. For example, in southern Africa, the largest known contiguous elephant population ranges across five countries: Angola, Namibia, Zambia, Botswana and Zimbabwe (Figure 7.7). This population is estimated at between 100 000 and 200 000, covering an area of approximately 700 000 km^2. Thus the protection of these elephants is a regional initiative as well as a part of the national elephant policy for each country.

In 1989, prior to the international ban on trade in ivory, one of the greatest concerns was the increasing isolation of elephant populations and the skewed demography of these herds following selective poaching. The demographic, genetic and ecological viability of small populations was, and still is, of grave concern. The AED can help to identify small, isolated areas of habitat where elephants exist and to target threatened elephant populations for increased conservation efforts.

Following the trade ban and the accompanying increase in security, dense populations of elephants in protected areas are coming into conflict with humans as they begin to range outside these protected areas.

This has been observed in many parts of East, Central and West Africa. According to Tchamba (pers. comm., 1992), human population pressure for land is now the primary problem of elephant management in northern Cameroun. With good data at the appropriate resolution, it would be possible to develop predictive models to estimate recovery rates of elephant populations under a variety of conditions. This would enable wildlife managers to look for strategies that permit humans and elephants to coexist on a sustainable basis. Long-term conservation and management needs for elephants will only be met if the problems of the growing human population are resolved.

The decline of the elephant in the past decade or so has primarily been due to intense poaching. However the range of the elephant also declines with the decline of the ecosystems upon which it depends and with the movement of humans into these areas. The AED allows the user to determine the degree of protection for different vegetation types within the range of the elephant (Douglas-Hamilton *et al.*, 1992). One can thus determine whether the different vegetation types are adequately covered in the existing protected area network. Of course, protection does not automatically come with the designation of an area as a park or game reserve. The effective management of these areas remains the greatest conservation challenge in most African countries. This requires commitment, training, education and adequate budgets to implement agreed strategies and management plans.

7.5 Future prospects of the AED

GIS technology, with its ability to manipulate, integrate and analyze spatial data, is a powerful tool for wildlife managers facing issues of ever-increasing complexity. However, as with any tool, GIS requires wise use in order to produce beneficial results that can then be implemented.

7.5.1 GIS specifications

The usefulness of the AED system must now be evaluated in terms of its actual and potential applications. No available GIS completely fulfills all the needs of its users; the AED is no exception. Moreover the variety, quality and price of computer hardware and software on the market is broad and expanding. The users are also evolving in terms of their increased awareness of GIS systems and the potential of such systems to help them address wildlife planning and management issues. Matching a system to defined needs remains crucial in implementing GIS technology. There is no ideal set of specifications for a GIS. Organizations combine different specifications to different degrees to suit their individual needs. Diverse applications inherent in the disciplines of planning and wildlife management require flexibility. We must reassess the systems that we are using in order to answer questions such as:

• Do we need a sophisticated system, with all

its accompanying complexities and expense to fulfill the needs of the users of the AED?

- Would the adoption of a simpler and cheaper system provide more users, both current and potential, with the opportunity for greater interaction with the AED?
- Will these changes provide a more accurate and current picture than would otherwise be the case?

(a) Data transfer

The interchange of data between the leading Geographical Information Systems is now relatively straightforward. Data are no longer 'trapped' in the originally adopted system(s). It should now be possible to increase the availability of elephant information and permit greater access to and interaction with the stored data.

(b) Legal status of the GIS data

The question of ownership or of custodianship of a database is a critical one. In certain countries, the legal status of geographic information is regulated under copyright law. However, the whole field of GIS is still in its infancy with regard to these and other aspects of commercialization.

The AED was established pragmatically to meet the needs of particular projects, users or organizations. This database is supported by the scientific input from the IUCN/SSC African Elephant Specialist Group which underpins the value and continuation of the AED. It is inevitable, however, that there are differences of opinion among users when dealing with issues such as data interpretation and representation. With whom does the ultimate responsibility of the reliability of the database rest? Should the responsibility be with the scientists who feed the database, the managers and technicians who drive the database, the organizations who house the database and disseminate the data, or the donors who finance the database? In order to truly address the problems of the future, all those supporting the database – whether scientists, planners, managers or sponsors – must achieve consensus with regard to these and other critical issues.

7.5.2 Database requirements and development strategy

The dearth of current data continues to be a problem as data constitute the backbone of an operating GIS. There is still a serious need for baseline surveys in many areas. The value of the AED ultimately depends upon greater emphasis being placed on the efficient implementation and successful outcome of field operations and surveys to acquire new data. With this in mind, it is important that a long-term strategy for the AED be developed. This plan should include appropriate methods to enable frequent changes to be made as new data are received and according to user needs.

7.6 Summary

The decade leading to the international ban on the ivory trade in 1989 witnessed a halving in the wild population of the African elephant, *Loxodonta africana*. Poaching for ivory, the destruction of suitable habitat, and human population expansion are all factors that have contributed to the decline of this species. The African Elephant Database arose because of a growing need by conservationists and development agencies alike for sound and reliable elephant data for both development and conservation planning.

The AED was established in 1987 in an attempt to ascertain the status of this species. This database is stored and manipulated within a GIS that holds data related to elephant numbers and distribution. Both locational and descriptive data are linked within the GIS, which can then be queried and analyzed to determine the temporal and spatial characteristics of the stored elephant data.

For the past six years, the AED has served as the primary foundation for a scientific consensus on the status of this species. This database, regularly updated as new data are received, gives timely and consolidated assessments of elephant populations for the biennial Convention on International Trade in Endangered Species (CITES). Recent database developments are enabling users to deal with a broader range of wildlife planning and management issues at regional, national and local levels. The AED is becoming an important tool for the identification of conservation priorities at the tactical, strategic and policy levels (Norton-Griffiths *et al.*, 1991). In this chapter two case studies have been presented that illustrate the broader potential of the AED.

Acknowledgements

Numerous organizations and individuals have been involved in the monitoring of elephants and the gathering of essential data over the years. Their contributions, whether in the form of raw data or participations in African Elephant Specialist Group (AESG) meetings to review the data, were invaluable in ensuring the value and reliability of this database. It is not possible to mention them all by name, but they include the members of the IUCN/SSC African Elephant Specialist Group (AESG), African government officials, and numerous individuals from all regions of the continent. As one of the founders of the database in 1986, and for a lifetime's dedication to elephants and to the collection of elephant information, Iain Douglas-Hamilton is surely the leader. Without the organizational context and technical support of GEMS and GRID of UNEP, and the unwavering encouragement of their respective directors – Dr Michael Gwynne and Dr Harvey Croze – there would not be a continental GIS for the African elephant. Finally, as manager of the AED, I should like to thank all those field workers, scientists, planners and managers who play a vital role through the provision of data, ideas and synthesis.

The AED has received major financial support from the United Nations Environment Programme, the Commission of the European Communities (EEC) and the Elsa Wild Animal Appeal.

Dr Richard Barnes, Dr Ruth Chunge, Dr Harvey Croze, Dr Iain Douglas-Hamilton, Dr Holly Dublin, Dr Michael Gwynne and Dr Chris Thouless were kind enough to review this chapter. I am very grateful to them for their advice and comments.

References

Alers, M.P.T., Blom, A., Masunda, T. *et al.* (1992) Preliminary assessment of the status of the forest elephant in Zaire. *African Journal of Ecology*, 30.

Barnes, R.F.W. and Barnes, K.L. (1992) Estimating decay rates of elephant dung-piles in forest. *African Journal of Ecology*, 30.

Barnes, R.F.W. and Jensen, K.L. (1987) How to count elephants in forests. IUCN African Elephant and Rhino Specialist Group. *Technical Bulletin*, 1, 1–6.

Barnes, R.F.W., Blom, A., Alers, M.P.T. *et al.* (1992) Human activities and the distribution of elephants in the forests of Gabon. [MS]

Burrill, A. and Douglas-Hamilton, I. (1987) *African Elephant Database Project: Final Report*, UNEP/GRID, Nairobi.

Clarke, R. (1986) *The Handbook of Ecological Monitoring. A GEMS/UNEP Publication* (ed. Robin Clark), Clarendon Press, Oxford.

Cobb, S. (1989) *The Ivory Trade and the Future of the African Elephant*, Ivory Trade Review Group, International Development Centre, Oxford.

Cumming, D.H.M. and Jackson, P. (1984) The status and conservation of Africa's elephants and rhinos, in *Proceedings of the Joint Meeting of IUCN/SSC African Elephant and African Rhino Specialist Groups at Hwange Safari Lodge*, August 1981 (published 1984).

Douglas-Hamilton, I. (1979) *The African Elephant Survey and Conservation Programme*. Annual Report to WWF/NYZS/IUCN.

Douglas-Hamilton, I. and Douglas-Hamilton, O. (1992) *Battle for the Elephants*, Viking, New York.

Douglas-Hamilton, I., Michelmore, F. and Inamdar, A. (1992) *African Elephant Database*. Report to the EEC, February 1992. Published by UNEP/GEMS/GRID.

Fay, J.M. and Agnagna, M. (1991) Forest elephant populations in the Central African Republic and Congo. *Pachyderm*, **14**, 3–19.

McClanahan, T.R. (1986) Quick population survey method using faecal droppings and a steady state assumption. *African Journal of Ecology*, **24**, 37–9.

MacKinnon, J. and MacKinnon, K. (1986) Review of the protected areas system in the Afrotropical realm. IUCN/UNEP, Gland, Switzerland, 259pp.

Merz, G. (1986) Counting elephants (*Loxodonta africana cyclotis*) in tropical rain forests with particular reference to the Tai National Park, Ivory Coast. *African Journal of Ecology*, **24**, 61–8.

Michelmore, F. (1991) *The African Elephant Database: A Technical Report*. United Nations Environment Programme. Global Resource Information Database Case Study Series No.5: 1–173.

Michelmore, F., Beardsley, K., Barnes, R.F.W. and Douglas-Hamilton I.(1989) Elephant population estimates for the Central African countries, in *The Ivory Trade and the Future of the African Elephant* (ed. Stephen Cobb),International Development Center, Oxford.

Lamprey, R.H., Michelmore, F. and Lamprey, H.F. (1991). Changes in the boundary of the montane rainforest on Mount Kilimanjaro between 1958 and 1987, in *The Conservation of Mount Kilimanjaro* (ed. William D. Newmark), IUCN, Gland, Switzerland and Cambridge, UK, pp. 9–15.

Norton-Griffiths, M., Campbell, K. and Michelmore, F. (1991) Applications for Geographic Information Systems in wildlife management, in *Wildlife Research for Sustainable Development*. Proceedings of an International Conference held in Nairobi, Kenya, by the Kenya Agricultural Research Institute (KARI), the Kenya Wildlife Service (KWS) and the National Museums of Kenya (NMK), April 22–26, 1990.

Short, J. (1983) Density of seasonal movements of forest elephant (*Loxodonta africana cyclotis* Matschie) in Bia National Park, Ghana. *African Journal of Ecology*, 21, 175–84.

Tchamba, M.N. (1992) Defaecation by the African forest elephant (*Loxodonta africana cyclotis*) in the Santchou Reserve, Cameroun. *Nature et Faune*, 7, 27–31.

Thouless, C. and Dyer, A. (1992) Radio-tracking of Elephants in Laikipia District, Kenya. *Pachyderm*, **15**, 34–9.

Wing, L.D. and Buss, I.O. (1970) Elephants and forest. *Wildlife Monographs*, **19**, 1–92.

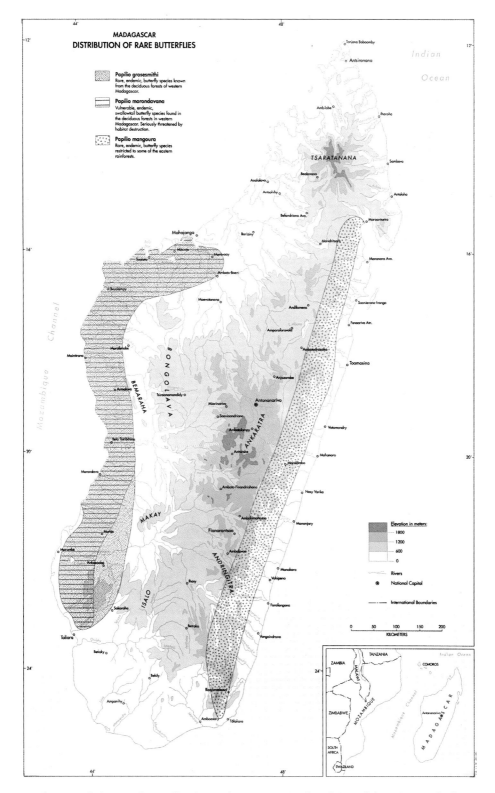

PLATE 1 Distribution of the rare butterflies in Madagascar (reproduced in collaboration with the Cartography Division of the World Bank).

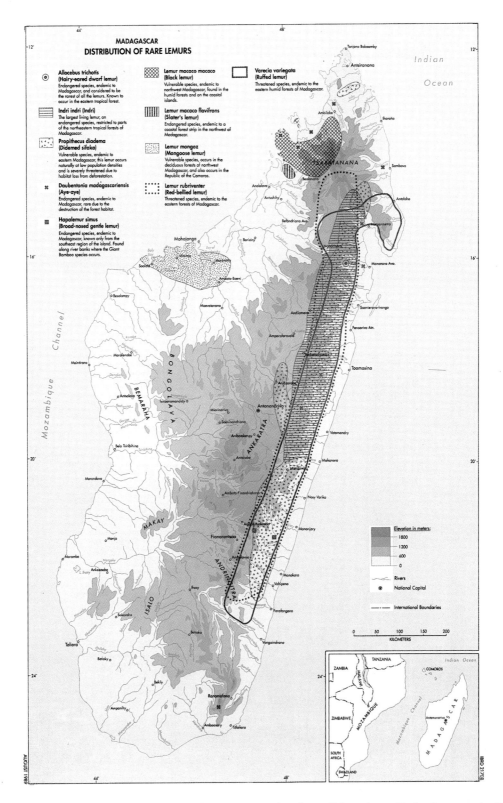

PLATE 2 Distribution of the rare lemurs in Madagascar (reproduced in collaboration with the Cartography Division of the World Bank).

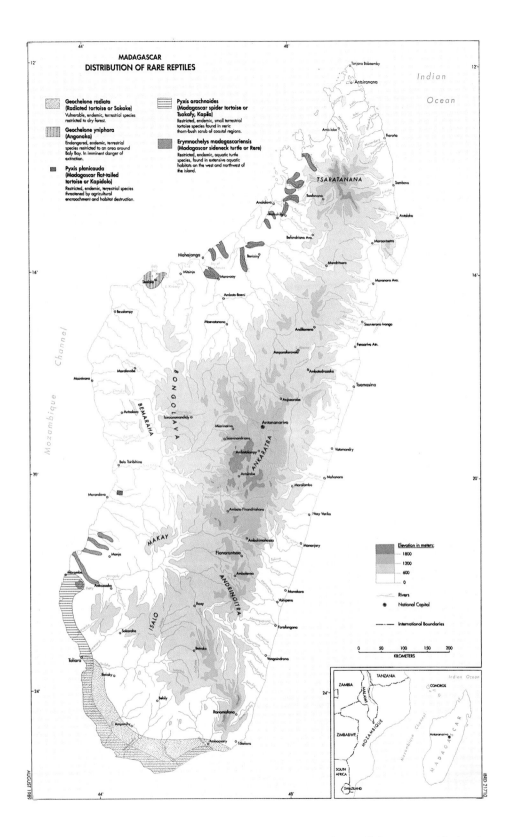

PLATE 3 Distribution of the rare reptiles in Madagascar (reproduced in collaboration with the Cartography Division of the World Bank).

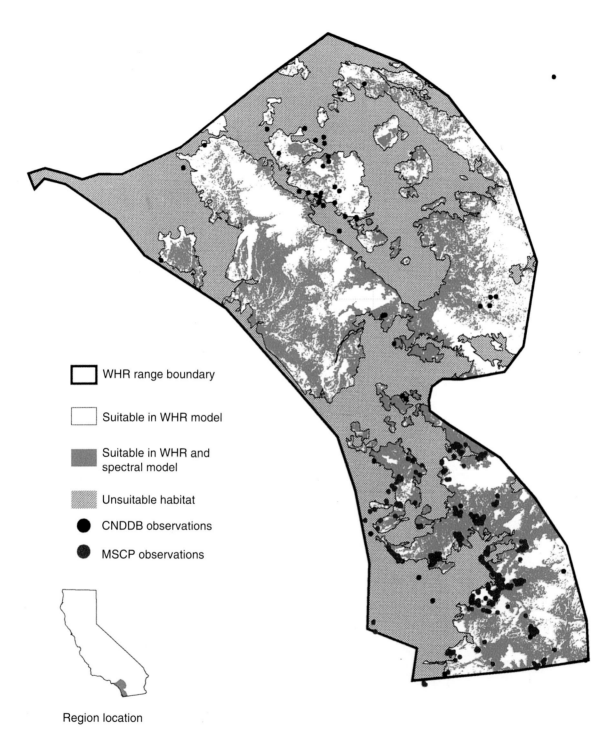

WHR range boundary

Suitable in WHR model

Suitable in WHR and spectral model

Unsuitable habitat

● **CNDDB observations**

● **MSCP observations**

Region location

PLATE 4 Whiptail distribution within the range limits. Areas not shaded in tan are suitable according to the WHR model; superimposed in green are pixels identified in spectral classification as likely to contain whiptail observations. Also superimposed are point observation data from both the MSCP and CNDDB datasets.

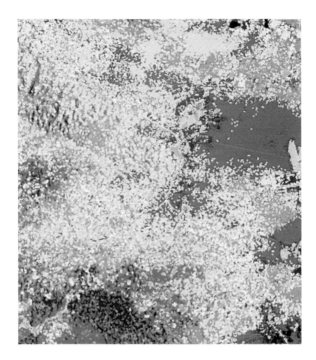

PLATE 5 Photo representation of the unsupervised classification produced from a LANDSAT subscene in southern Belize. This image was recorded on 25 March 1987.

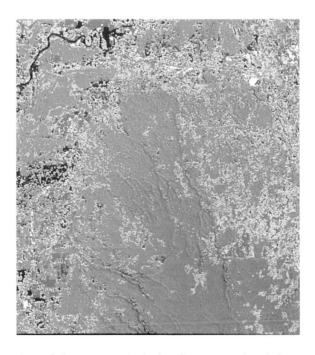

PLATE 6 Photo representation of the unsupervised classification produced from a LANDSAT subscene in northeastern Costa Rica. This image was recorded on 6 February 1987.

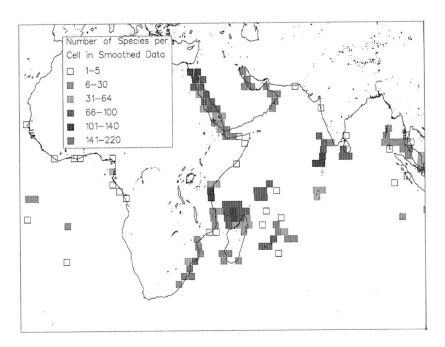

PLATE 7 Species density in the Indian Ocean for the all-families sample. Note the clusters around Madagascar and East Africa, the Red Sea, and the Laccadive–Maldive–Chagos–Sri Lankian regions. Sampling is weak on western Sumatra.

(a)

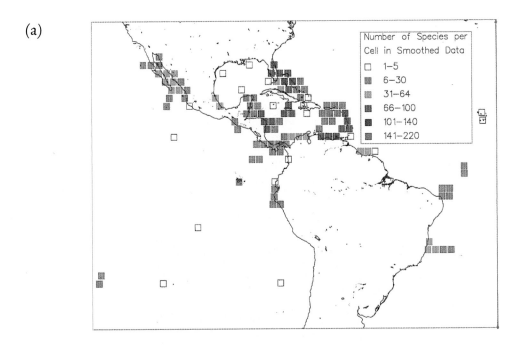

PLATE 8(a) Species density in the eastern Pacific and the western Atlantic.

(b)

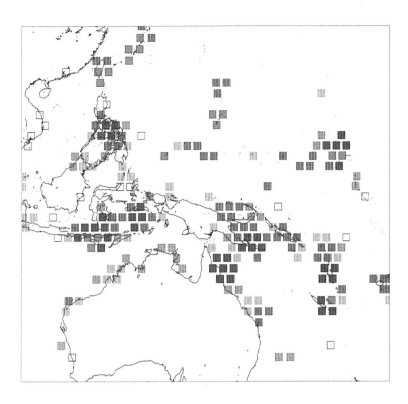

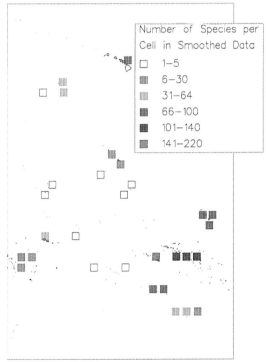

PLATE 8(b) Species density in the Indo–Australian Archipelago located in the western and central Pacific. Note the high values in the eastern end of the Archipelago. Relatively high values in the Ryukyus and Palau indicate the influence of better sampling in this region.

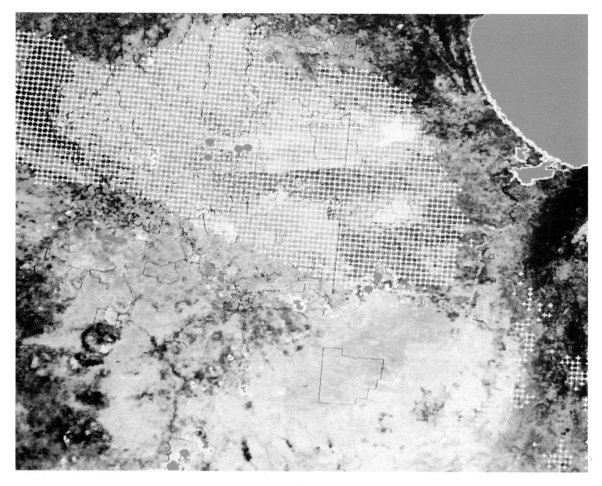

PLATE 9 BIOCLIM prediction for Malleefowl (*Leipoa ocellata*) in southern Australia. The white crosses show areas of predicted suitable climate in the central part of southern Australia. The predictions are based on known localities of the Malleefowl (shown as green dots) which were obtained from the Australian Nature Conservation Agency database of Bird and Bat Banding records. The background consists of three Normalized Difference Vegetation Index (NDVI) data sets covering late summer, winter and spring derived from NOAA AVHRR satellite data. The NDVI values have been inverted so that areas of consistently high NDVI (i.e. generally green all the year) are dark, while the more sparsely covered habitat would be expected to occur in the darker gray areas (i.e. areas of mallee *Eucalyptus* species). Areas outlined in dark blue are areas of Crown land and include biosphere reserves and national parks.

Chapter eight

Designing protected areas for giant pandas in China

John MacKinnon and Robert De Wulf

8.1 Decline of the giant panda

The familiar black and white giant panda is famous throughout China and the world as a symbol of the plight of endangered species. Fossil remains show that the giant panda was formerly distributed over a much larger area than today, reaching Burma, northern Vietnam, and much of eastern and southern China. Historical records in China also demonstrate that it was formerly much more widespread than today. Figure 8.1 shows this declining range of the giant panda.

However, as a result of hunting, logging and clearing of forest for agriculture, the range of the giant panda is now condensed to just six isolated mountain ranges in Sichuan, Gansu and Shaanxi provinces. These remaining areas cover 29 500 km².

The People's Republic of China take the con-servation of this natural treasure very seriously and the first four panda reserves were established in 1963. Today there are 13 panda reserves with a total area of 5827 km² (Figure 8.2). These reserves protect not only the giant panda but also many other rare species such as the red panda, the golden monkey, the takin, and a great wealth of plant species and important water catchment areas.

Despite the establishment of these reserves and the passing of strict laws to protect the giant panda (i.e. there is a death penalty for killing or trading in pandas), their numbers have continued to decline steadily over the last 25 years. Two major surveys were conducted to assess the entire panda distribution. Between 1974 and 1975 surveys were directed by Chinese teams and again between 1985 and 1988 they were guided by a joint China/WWF team. These surveys were done together with

Mapping the Diversity of Nature. Edited by Ronald I. Miller.
Published in 1994 by Chapman & Hall, London. ISBN 0 412 45510 2.

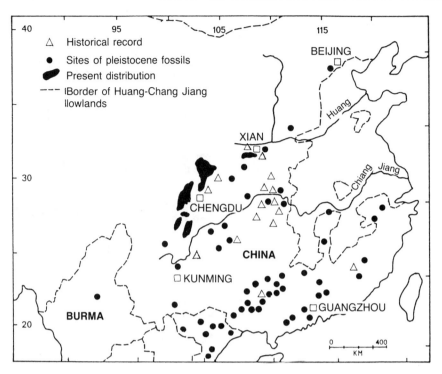

FIGURE 8.1 This map shows the shrinking range of the giant panda from prehistoric times to the present day (after Schaller *et al.*, 1985).

detailed mapping of forest cover derived from satellite imagery data, and the results show that the area of suitable habitat occupied by giant pandas in Sichuan has shrunk from a figure of over 20 000 km² in 1974 to only 10 000 km² in 1989. This represents a loss of about half the area in only 15 years. The situation in Gansu and Shaanxi is similar.

Another reason for the panda's declining numbers is that large areas of arrow bamboo, a mainstay of the panda's diet, died after mass-flowering since 1974. When the arrow bamboo died, many pandas starved or migrated. In the worst affected areas up to 80% of pandas were thought to be lost. The bamboo flowers at 40–60 year intervals and requires 15 to 20 years to regenerate to a size that pandas can eat. Therefore many giant pandas have now suffered over a long period from malnutrition. A second mass-flowering in 1983–4 killed the

bamboo areas not affected by the first flowering. The effects of these two natural events were worsened because farmers had cleared valleys into which pandas could have migrated to find secondary bamboo species.

Between 1986 and 1989 a special management plan was prepared jointly by the Ministry of Forestry and WWF to save the giant panda (MacKinnon *et al.*, 1989). This plan discusses management measures and develops proposals for protecting and maintaining a healthy population of wild giant pandas. This plan extends the reserve system, restores damaged habitat, and proposes a management scheme for the remaining panda habitat both within and outside panda reserves.

Use of satellite imagery from the habitat of the giant panda proved to be of vital importance for understanding the crux of the real problems affecting the giant panda. These

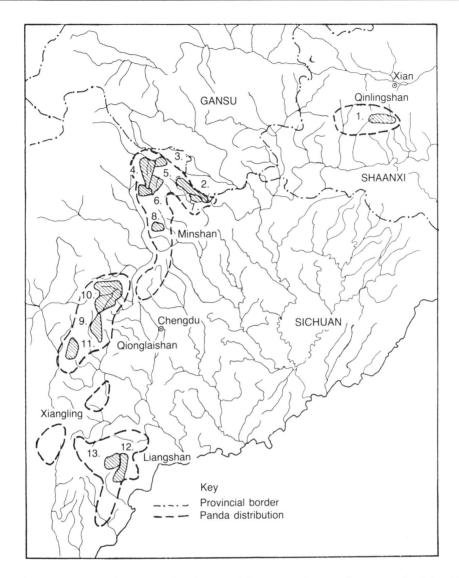

FIGURE 8.2 This map presents the current distribution of the extant giant panda reserves in China. The displayed panda reserves include: 1, Foping; 2, Baishuijiang; 3, Baihe; 4, Juizhaigou; 5, Wanglang; 6, Huanglongsi; 7, Tangjiahe; 8, Xiaozhaizigou; 9, Fengtongzhai; 10, Wolong; 11, Labahe; 12, Dafengding Mabian; 13, Dafengding Meigu.

imagery data also served to focus the viewpoints of the Chinese and foreign scientists involved in the planning and identification of the management solutions needed to save the giant panda.

8.2 Traditional view of the problem

Pandas were seen to die whenever the bamboo flowered and died. At such times starving pandas would wander into village lands. Villagers were encouraged to rescue such animals, feed them back to health and bring them into captivity. The Chinese government built special holding compounds for these rescued animals and zoos increased their capacity to hold pandas. Great emphasis was placed on attempts to breed pandas in captivity and thus to save the species *ex situ*.

The great problem of *in situ* conservation was manifest in the awkward behaviour of bamboo. Efforts to study bamboo were established to see if its flowering could be artificially controlled. A number of reasonable solutions to this problem were available. One possibility was to stagger the flowering of one or more of the important bamboo species. A second solution was to introduce alternative bamboo species that would supply the pandas with food while the critical bamboo was subject to flowering periods. A third solution was to accelerate the regeneration of the critical bamboo species subsequent to flowering. All of these options needed to be considered.

We now realize that any attempt to manipulate either the natural environment of the bamboo and/or to change the natural cycles of bamboo flowering would have been a grave management error. Also any further attempt to rescue more pandas would have increased the extinction probability of the giant panda in the wild. Any of these potential strategies would have harmed the status of this species. All of these mistakes were averted, thanks largely to the relatively new remote-sensing technology.

8.3 Factors threatening the giant panda

Two time series of satellite images showed the reduction and rapid fragmentation of the giant panda's habitat. More than any other factor it was this perspective provided by satellite imagery that changed the Chinese managers' views about the main threat to panda survival.

The pattern of localities where pandas had

become extirpated between 1974 and 1989 was not related to bamboo flowering (Figure 8.3). Pandas survived, albeit at reduced density, in every area affected by bamboo flowering. But the pandas vanished from several isolated and fragmented forests at the edges of their distribution. In total, the habitat extent of the species was reduced by 50% during these 15 years. We now know that the panda's problem during this period was human encroachment, fragmentation and an edge effect.

Pandas were disappearing along the interface with humans as a result of trapping (e.g. sometimes they were accidentally caught in musk deer snares) and shooting. In one operation the Sichuan provincial authorities rounded up a total of 146 illegal panda skins. In other cases during this period, four persons were shot for killing giant pandas or smuggling panda skins. A combination of human encroachment and the edge effect, framed most poignantly by the impact of hunting, produced a severe reduction in the population of giant pandas.

8.3.1 Threat to bamboo regeneration

It soon became clearer that manipulating the bamboo flowering cycles was not only unnecessary but could even be hazardous to the survival of the bamboo, the obligate food resource for the panda. Synchronous mass flowering of bamboos has evolved over a long period, it is clearly adaptive, and it may be essential for the well-being and survival of these plants. Any artificial interference with the timing or synchrony of flowering may pose a threat to the bamboo and ultimately to the pandas that depend on it for food. The importance of bamboo flowering gradually dropped in the list of priority threats to the pandas.

Currently the bamboos are regenerating well and the immediate threat of starvation is past. Panda populations are expected to recover naturally as long as they are adequately pro-

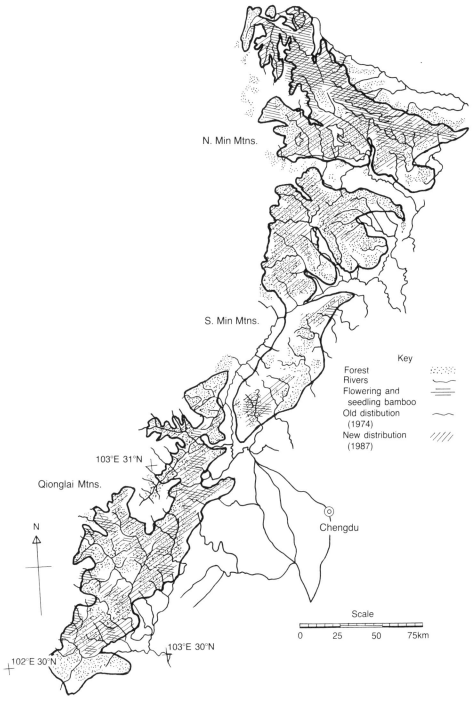

Key

Forest
Rivers
Flowering and
  seedling bamboo
Old distibution
  (1974)
New distribution
  (1987)

N. Min Mtns.

S. Min Mtns.

Qionglai Mtns.

Chengdu

103°E 31°N

102°E 30°N

103°E 30°N

N

Scale

0   25   50   75km

FIGURE 8.3 This map shows the distribution of giant pandas in the 1974 field survey in contrast to the panda distribution recorded in the 1987 field survey. The distribution of bamboo flowering and forest cover patterns is also displayed here for comparative purposes.

tected from further habitat loss and poaching. Additional research into bamboo biology is still important. The restocking of lands damaged during this episode should proceed, but widespread application of bamboo management is not now a priority and it could be dangerous.

8.3.2 Problems of habitat fragmentation

During the 1970s and 1980s many starving pandas were observed wandering into village lands. These animals were thought to be desperately famished pandas seeking human help. The capture and captive breeding of these animals was then seen as the principal solution to this problem.

It is now recognized that these are hungry pandas crossing cultivated lands to reach other 'food rich' areas of unflowered bamboo habitat. The policy of rescuing such animals into captivity is not in the best interest for maintaining panda populations in the wild. Today it is still important to pursue improved techniques for captive breeding, but only as a secondary tactic for saving the species. New methods to introduce captive bred animals back into the wild are needed to reduce the present size of the large captive population.

Priorities for saving the panda needed to become redirected. The geometry of the areas of remaining protected habitat needed strengthening. A new focus on maintaining and even recreating links or corridors between semi-isolated populations was needed. This called for concentrating efforts on the weak points in the populations rather than on the main reserves and population strongholds. The broader perspective required by these tasks was provided by analysis of satellite images.

8.4 Use of satellite imagery

Developing a management plan to save the giant panda required the survey of an area of 30 000 km². The quickest and most cost-effective way to evaluate habitat condition was to use satellite imagery. These data were then combined with the results of a ground survey to produce high-quality maps of panda habitat. A team of 30 people visited panda habitat in over 30 Chinese counties and they collected field data for over 2000 localities.

For habitat classification, Landsat MSS imagery is limited due to its coarse resolution. Nevertheless, Landsat MSS imagery was chosen because it offered the best-known trade-off between cost and accuracy for an up-to-date multispectral synoptic coverage. Another advantage of Landsat was the archive of images taken between 1975 and 1988 that enabled detection of change in forest cover over time (see Appendix 8.1 for a listing of the Landsat frames used in this analysis).

A general map was prepared using contact prints of IR bands from 1983 images at the 1:1 000 000 scale. Due to limitations of budget and time, no digital processing was attempted. Imagery from 1975 was also purchased to examine the pattern of forest loss over the preceding decade. Once produced, this map allowed us to locate both areas of resource conflict and blocks of panda habitat in danger of becoming isolated.

Mountain barriers were easily identified in the imagery as ridges of permanent snow or bare rock. Coniferous forests (i.e. dark to blackish gray tones), deciduous forests (i.e. light gray tones), and cleared, formerly forested areas (i.e. mottled middle gray tones) could be distinguished easily in the images. However, the current condition of bamboo could not be identified even from air photos (Morain, 1986) and therefore data describing bamboo quality needed to be based on ground surveys.

The locations of the occupied panda ranges were obtained from the ground surveys. When the satellite imagery and the ground survey data were overlaid, a serious fragmentation of the panda distribution soon became apparent. The composite map revealed about 30 extant population fragments and it also showed that

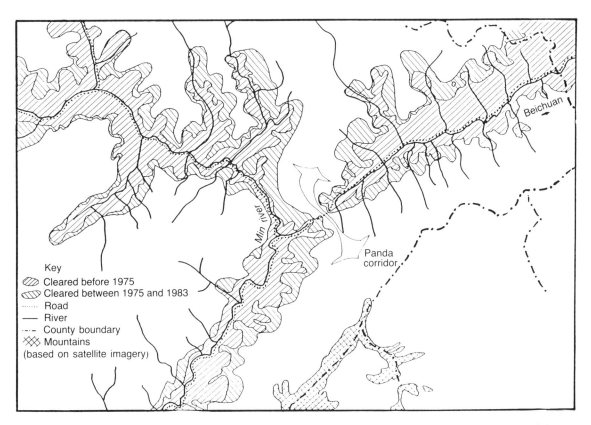

FIGURE 8.4 This map depicts a division occurring in the giant panda population as a result of human encroachment. This is an example of a panda corridor located between the Min river and the Beichuan road that is being severed.

several additional areas were in danger of becoming isolated (Figure 8.4). Moreover it was obvious that in the period between the two ground surveys (1974–5 → 1986–8), many changes had occurred. Pandas had disappeared from a group of areas that were either small and isolated (e.g. Figure 8.5), at the edge of the species distribution, or where the habitat was devastated. It was apparent that habitat loss and fragmentation was the most serious threat to the giant panda.

The map produced using the satellite imagery thus formed the basis of the species management plan and permitted a redesign of the panda reserve system. Within this system, production forestry and human encroachment

were halted and reforestation was suggested in areas further threatened by fragmentation. Furthermore, to re-establish links between semi-isolated subpopulations of giant pandas, 15 critical corridors in need of special protection were identified. A more detailed mapping of habitat within reserves, based on enlargements at a scale of 1:200 000, assisted in the production of management plans. The final system included plans for zoning and relocation of villages in areas of panda reserves.

8.5 Redesign of the reserve system

Many extant giant panda reserves are too small or too isolated to maintain populations over

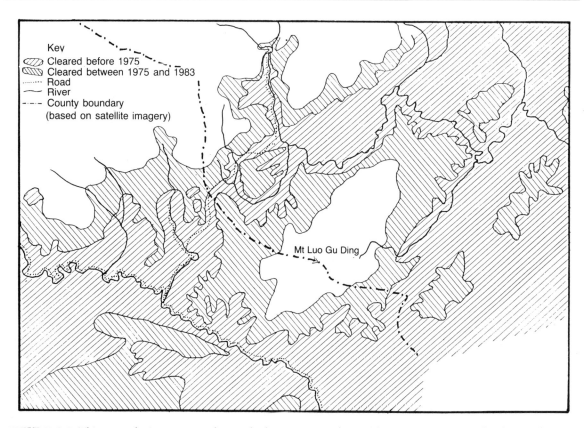

Key
- ⬚ Cleared before 1975
- ⬚ Cleared between 1975 and 1983
- ······ Road
- —— River
- ·—·— County boundary
  (based on satellite imagery)

Mt Luo Gu Ding

FIGURE 8.5 This map depicts an area from which a giant panda population was extirpated. The pandas were resident on Mt Luo Gu Ding but they were extirpated soon after they were isolated from the main population. This extirpation occurred between 1976 and 1978. This map was produced using satellite imagery data.

the long term. Consequently, it was necessary to extend and redesign the present reserve system. To date, the boundaries of several reserves have been revised. In some cases it became apparent that it would be quite difficult to resettle the human populations. Therefore the new park boundaries were drawn to exclude these people. In other cases extant reserve boundaries were extended to establish corridor links with adjacent panda populations.

At present, 14 new reserves are proposed with a total area of 3302 km². These new reserves will serve several purposes: they will link existing reserves and will protect other important panda populations not currently included in the reserve system. The revised system of reserves will double the number of pandas contemporaneously living within reserve boundaries. The approved changes to the giant panda reserve system in Sichuan and Shaanxi provinces are presented in Figure 8.6.

8.6 Identification of critical habitat corridors

Where possible, the existing panda reserves should be connected, and once established, these panda habitat links should be carefully preserved. In some cases it will even be necessary to rebuild suitable habitat corridors to

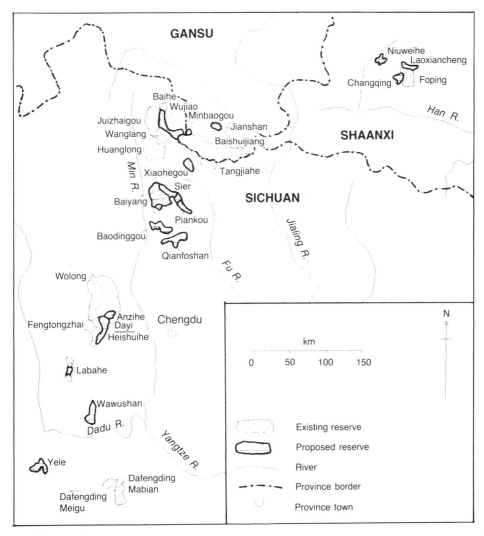

FIGURE 8.6 This map shows the currently approved extensions to the giant panda reserve system in China. This represents the global distribution of giant pandas.

maintain genetic exchange between fragmented subpopulations. New regulations should facilitate activities that help to maintain natural conditions in and around reserves. These activities will include the restoration of damaged land in nature reserves, the relocation of villages impinging upon panda habitat, and the modification of forestry procedures in panda habitat outside of reserves.

A total of 15 corridors are identified as critical for maintaining connections between panda reserves and to ensure that genetic exchange occurs between isolated panda populations. These corridors require strict protection. Seven of the areas require replanting to re-establish corridors of suitable habitat for pandas. These replanted corridors should be at least 1 km wide and they should bridge the

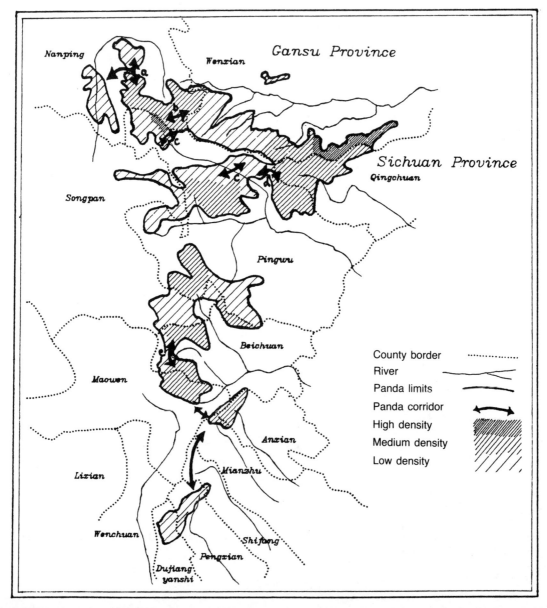

FIGURE 8.7 This map depicts the density distribution of the giant panda in the Minshan area of China at a large map scale. The location of the critical corridors identified in the Min mountain area of Sichuan province are shown.

narrowest and flattest gaps between the panda habitat blocks. The corridors should be replanted with native tree and bamboo species. All of the critical corridors should be strictly prohibited to logging, gathering and grazing. The locations of the critical corridors in the Minshan Mountains of Sichuan are depicted in Figure 8.7.

8.7 Management planning inside reserves

Satellite imagery combined with topographical and field data can be used to create detailed vegetation maps of individual reserves. A vegetation map produced for the Wolong Nature Reserve, Sichuan, is presented in Figure 8.8. This map was subsequently used to design a zoning scheme for this reserve.

Use of remote-sensing technology enables panda management to occur with a minimum movement of people and only in areas that are essential for panda survival. Relocation of human settlements is a difficult, expensive and sensitive measure. To resettle villages it is necessary to pay compensation, to locate alternative agricultural areas, and to obtain the agreement of the inhabitants. It is also clear that such relocation is easier with ethnic Han Chinese than with minority groups. For example, in the Tangjiahe reserve, all of these criteria were met and the villagers were successfully moved out of the reserve. In the Wolong Nature Reserve, one of the critical zoning problems is the removal of villagers from the core area, but so far efforts to persuade villagers of the Wolong core area to relocate have failed. Partly this is due to the authorities being less strict with minority groups (Wolong) as opposed to Han Chinese in Tangjiahe.

Positive restoration measures need to be undertaken where panda habitat is destroyed inside reserves as a result of previous agriculture and forestry activities. These measures include the planting of an appropriate mix of local trees and bamboo species and the rehabilitation of eroded mountain slopes in critical areas. Exotic bamboo, even species taken from other parts of China, *should not* be introduced into these ecosystems. Once again, remotely sensed imagery proved invaluable for the mapping of the exact areas to be planted. Figure 8.9 shows one planned corridor inside the Wolong Nature Reserve.

8.8 Satellite imagery interpretation used for the Chinese biodiversity database

Satellite imagery can be used to map wildlife habitat over broad (even continental scale) areas and monitor the fate of such habitat over time. This can enable the wildlife manager to indirectly monitor the fate of hundreds of species associated with different types of habitat (MacKinnon, 1993). Some of the scale and resolution problems associated with mapping habitats at this broad scale in different parts of the world are presented in Chapters 3, 5, 7, 9, 10 and 12.

Analysis of panda habitat in China requires input of the maps and data into a GIS. Examination of 520 Landsat frames is required to map all the forest and natural vegetation of China. Another thematic data layer in the GIS is a map of potential habitat that is based upon the vegetation map of China (*People's Republic of China Vegetation Map*, Scale 1:4 000 000, Chinese Academy of Sciences, Beijing, 1982). A most important part of the GIS is the map of all the nature reserves in China (over 600 sites). Each digital map comprises a separate layer and all of these maps are then overlaid for comparative analysis purposes.

The assembled database maintains the compiled data from the results of the map analyses. The status of every vegetation type in each province in China is listed according to whether it is an original and/or a remnant area and whether it is currently protected or within a proposed protected area. One file contains a variety of species-related data. These include: the vegetation types from which a species is known, the biogeographic regions with which a species is linked, the altitudinal range for a species, the known density occurrences of the species, and threats to the species (e.g. hunting, trade, etc.). Another file is a compilation of data regarding the proportion of surveyed areas where a species is theoretically expected in

137

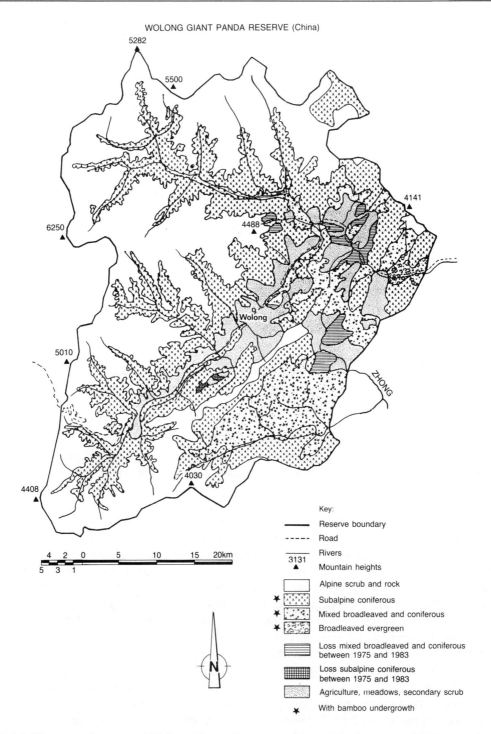

FIGURE 8.8 The vegetation map of Wolong Nature Reserve used in zoning the reserve. This map is based upon satellite images from Landsat 1975 (frames 140/38–39) and Landsat 1983 (frames 130/38–39).

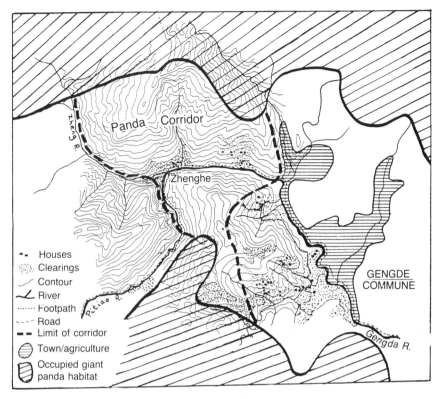

FIGURE 8.9 A critical panda corridor located in the Wolong reserve. The area was identified from remote sensing, use of topographical maps (to identify areas of less steep terrain) and ground checking. It is planned to be protected from further encroachment and reforested where necessary.

relation to the areas where a species is confirmed to be present. This proportion monitors the success of the map overlay analysis procedures.

The relational database is able to evaluate this information and give a crude conservation status report on any entered species, including an estimate of total population (see Appendix 8.2 for the giant panda status report). Thus the MASS database functions as an early warning system to identify endangered species issues before the problem is recognized from the available field data. The database predicts an estimate of about 1200 giant pandas remaining in the wild. Using conventional field methods, this same figure required a two-year field survey with 30 people climbing steep mountains and enduring considerable hardship.

MASS is currently being linked to an Arcad GIS system so that the overlays and measurements can be automated and species output can include graphical data presentation. The program links to a bibliography file which produces a relevant bibliography for each species. Another interesting point is that the program predicted E Mei Shan as a possible location for giant pandas. At the end of last year a giant panda was found and trapped on E Mei Shan. The first record ever of the species at that locality.

Acknowledgements

The authors would like to thank: the World Wide Fund for Nature who sponsored the studies referred to in this paper; the Ministry of Forestry, Beijing; Bureau of Forestry, Sichuan Province; and Management Bureau of Wolong National Nature Reserve for their assistance and cooperation.

Appendix 8.1

This is a list of Landsat MSS images used in this analysis. The frames are standard Landsat views, the date refers to date that the image was recorded, and the spectral bands refer to the image wavebands purchased for analysis. For the spectral bands: 2=red, 4=near infrared, 7=near infrared.

Landsat frame #	Recording date	Spectral band
139/37	04/10/75	7
139/38	04/10/75	7
140/38	17/06/75	7
140/39	17/06/75	7
129/37	12/05/83	4
129/38	12/05/83	4
130/38	19/05/83	2,4
130/39	19/05/83	2,4

Appendix 8.2

This is a summary printout of the MASS data regarding the giant panda. Entries under bio-units and habitats are codes used in the database for China.

Status summary for *Ailuropoda melanoleuca* (Giant panda)
Distributed in biounits: 39d, 39e, 15d
Occupies habitats: TCF, CCF, CBM, DEB

Habitat summary:

% original habitat lost =	66.1
% original habitat protected =	9.8
% original habitat proposed =	4.3
% remaining habitat protected =	32.2
Expected protected population =	800
Expected total population =	1200
Based on entered density of =	0.20/sq.km
Reliability rating =	good
% age population protected =	67
Red Data Book status =	Endangered
Computer assigned status =	Vulnerable
Species listed on CITES appendix:	1
Species protected in countries:	China

Evaluation of habitat status (areas given in square km)

Unit	Admin div.	Veg. type	Origin. area	Remain. area	Rem. %	Prot. area	Prot. %	Prop. area	%
15d	SHX	TCF	2 081	0	0	0	0.0	0	0.0
15d	SHX	DEB	4 800	2 500	52	300	6.2	0	0.0
39e	GAN	CCF	8 150	0	0	1 870	22.9	1 590	19.5
39e	SZE	CCF	16 337	8 169	50	1 870	11.4	90	0.6
39e	SZE	DEB	664	133	20	0	0.0	0	0.0
39d	SZE	CCF	6 304	2 522	40	289	4.6	0	0.0
39d	SZE	DEB	5 640	1 692	30	0	0.0	0	0.0
Totals			43 976	15 016	34	4 329	9.8	1 680	4.3

Giant Panda is expected/confirmed in the following areas

Name	Country	Admin.	Total area (km²)	Confirmed	Status
Wolong	CH	SZE	2000	*	4
Wang Lang	CH	SZE	332	*	4
Tang Jia He	CH	SZE	300	*	4
Ma Bian Da Feng Ding	CH	SZE	340	*	4
Mei Gu Da Feng Ding	CH	SZE	180	*	4
Jiu Zhai Gou	CH	SZE	600	*	4
Feng Tong Zhai	CH	SZE	400	*	4
Bai He	CH	SZE	200	*	4
Xiao Zhai Zi Gou	CH	SZE	67	*	4
La Ba He	CH	SZE	330	*	4
Huang Long Si	CH	SZE	400	*	4
Tai Bai Shan	CH	SHX	563		4
Fo Ping	CH	SHX	292	*	4
Bai Shui Jiang	CH	GAN	1985	*	4
Tou Er San Tan	CH	GAN	319		4
Mai Cao Gou	CH	GAN	36		4
Hei He	CH	GAN	42		4
E Mei Shan	CH	SZE	100		4
An Zi He	CH	SZE	101	*	4
Da Yi	CH	SZE	0	*	4
Wa Shan	CH	SZE	228	*	4
Si Er	CH	SZE	190	*	4
Wu Jiao	CH	SZE	371	*	4
Xiao He Gou	CH	SZE	141	*	4
Bao Ding Gou	CH	SZE	196	*	4
Qian Fo Shan	CH	SZE	172	*	4
Niu Bei Liang	CH	SHX	165		4
Pian Kou	CH	SZE	197	*	4
Bai Yang	CH	SZE	583	*	4
Jian Shan	CH	GAN	40	*	4
Hong Qi	CH	SZE	250	*	4
Ma Mi Ze	CH	SZE	0	*	4
Zhi Bai Shan	CH	SHX	123		4
Yi Zi Ya	CH	SZE	250	*	4
La Zi Kou	CH	GAN	48		4
Zhang Qin	CH	SHX	298	*	4
Lao Xian Cheng			164		4

References

Campbell, J.J.N. and Zisheng, Q. (1983a) *Flowering and Population Dynamics of the Bamboo Stems in the Range of Giant Pandas, China.* Unpublished report to WWF/IUCN.

Campbell, J.J.N. and Zisheng, Q. (1983b) Interaction of giant pandas, bamboo and people. *Journal of the American Bamboo Society,* 4, 1–34.

Chinese Academy of Sciences (1982) *People's Republic of China Vegetation Map.* Scale 1:4 000 000 Beijing.

De Wulf, R.R., Goossens, R.E., MacKinnon, J.R. and Cai, W.S. (1988) Remote sensing for wildlife management: Giant panda habitat mapping from Landsat MSS images. *Geocarto International,* 1, 41–50.

De Wulf, R.R., Borry, F.C., De Roover B.P. and MacKinnon, J.R. (1990) Monitoring deforestation for nature conservation management purposes in Sichuan and Yunnan provinces, People's Republic of China, using multitemporal Landsat MSS and SPOT-1 HRV multispectral data, in *Proceedings of the 9th EARSeL Symposium,* Directorate General for Science

Research and Development Joint Research Centre – Ispra Site, pp. 317–23.

MacKinnon, J.R., Fengzhou, B., Mingjiang, Q. *et al.* (1989) *National Conservation Management Plan for the Giant Panda and its Habitat*, Ministry of Forestry of the People's Republic of China and WWF – World Wide Fund for Nature, Hong Kong, 157pp.

MacKinnon, J.R. (1993) *The Logic of MASS*, IUCN.

Morain, S.A. (1986) Surveying China's agricultural resources: patterns and progress from space. *Geocarto International*, 1, 15–24.

Schaller, G.B., Jinchu, H., Wenshi, P. and Jing, Z. (1985) *The Giant Pandas of Wolong*, Chicago University Press, 298pp.

Part Six

Mapping the Global Distributions of Species

Chapter 9 offers an approach developed in a biodiversity project at BirdLife International* for globally mapping the distribution patterns of restricted-range bird species. The database that is fully documented in this chapter was designed to store the full details of the available distributional records as well as habitat information for birds occurring in relatively poorly known tropical regions of the world. The information compiled was used to identify 221 Endemic Bird Areas (EBAs) that are key areas for biodiversity conservation. The EBAs can be used to help assess the effects of habitat destruction and modification on global biodiversity, to evaluate existing protected areas systems, and to identify areas for designation as new protected areas. The methodology developed at BirdLife can also be used to investigate the patterns in the distribution of more common species and help to assess their conservation status.

Coral reefs in many areas of the world are degraded by human activities, but geographic information about the fishes (approximately 4000 species) on those reefs is unavailable or widely scattered. Based upon an ongoing global study, Chapter 10 describes the use of a GIS to assemble, analyze and present information to support the conservation of coral reef fishes. This includes the geographic occurrence, threat status, and areas of high diversity and endemism ('hotspots') for these species.

A technique that uses essentially square, equal-area grid cells is introduced in Chapter 10 as a useful tool for the compilation, analysis, and presentation of data important to conservation assessment. The technique can be used to define and locate areas with high overall species richness, many endemic species, high taxon richness, intense human environmental impacts, or with a need for increased biotic sampling. The accuracy of the results using this technique will only be moderated by the level of taxonomic knowledge and the thoroughness of biological inventories. This technique will be extremely useful to planners and conservationists concerned with protecting the diversity of life in the oceans.

* As of March 1993, the International Council for Bird Preservation (ICBP) has operated under the new name of BirdLife International.

Chapter nine

Mapping the distributions of restricted-range birds to identify global conservation priorities

Michael J. Crosby

9.1 Introduction

The conservation of biodiversity has become one of the major environmental issues of the late twentieth century. It is predicted that the uncontrolled exploitation of natural resources can soon lead to a mass extinction of species (see, e.g., Myers, 1979; Wilson, 1988). In the past, the efforts of conservationists mainly focused upon a small number of species, typically large, charismatic and threatened animals. It is now clear that initiatives directed only toward these species cannot adequately address the more fundamental problems associated with general biodiversity loss.

Biodiversity is the total variety of life on earth. Our comprehension of this variety is limited, since only about 1.4 million of an estimated 5 to 30 million species on the earth are described (Wilson, 1988). How do we approach the conservation of this biodiversity when it is so poorly understood? Resources could be devoted to taxonomic research in order to rapidly advance our knowledge of all biodiversity (May, 1990), but, essential as a continuing role for taxonomy certainly is, the limited time available is such that any useful results would appear too late to solve the problem. It is clear that a strategy to conserve overall biodiversity must be based largely upon

Mapping the Diversity of Nature. Edited by Ronald I. Miller.
Published in 1994 by Chapman & Hall, London. ISBN 0 412 45510 2.

information that we already possess on the better known groups of plants and animals.

Biodiversity is not distributed evenly. Some parts of the world are much richer in species than others, and some places support concentrations of species which are found nowhere else. On the basis of vertebrates, swallowtail butterflies and higher plants, 12 'megadiversity' countries have been identified (McNeely *et al.*, 1988; Mittermeier, 1988). These countries by themselves account for up to 70% of the world's diversity of species. The botanical 'hotspots' analysis of Myers (1988, 1990) emphasized the importance of narrowly distributed species as well as high diversity, and resulted in the identification of 18 'hotspot' areas which support 20% of the world's known plant species in approximately 0.5% of the world's land surface. While interesting at the global level, however, neither of these approaches is based upon sufficiently detailed distributional information to enable a focus beyond the country or 'hotspot' level.

Many countries, principally in the tropics, contain large numbers of species with small geographical ranges. Concentrations of these restricted-range endemics are found on certain islands or island groups or in discrete areas of a particular habitat or grouping of habitats in continental regions. These 'habitat islands' are usually associated with a mountain range, river valley, coastal lowland strip or other geographical feature. Those species which are confined to a certain vegetation zone and unable to disperse through surrounding areas, are particularly vulnerable to habitat modification and destruction. The areas where restricted-range species are concentrated are therefore of particular conservation concern. They contain a significant proportion of all biodiversity concentrated into a relatively small portion of the world's land area, when habitat destruction can lead to mass extinctions of species.

9.2 Context of this study

The BirdLife Biodiversity Project (ICBP, 1992; Stattersfield *et al.*, in prep.) started in 1988, with the aim of identifying important areas for the conservation of biodiversity at the global level. Birds were used here as indicators to identify areas that support concentrations of restricted-range species, which are known here as Endemic Bird Areas (EBAs). An extensive literature review also investigated the congruence between the locations of EBAs and documented centers of endemism of other major plant and animal groups.

The methodology of this Biodiversity Project is based upon studies of the restricted-range bird species of sub-Saharan Africa by Hall and Moreau (1962) and of Colombia and Ecuador by Terborgh and Winter (1983). Terborgh and Winter covered all species with ranges estimated to be below 50 000 km^2. This definition was applied throughout the world by the Biodiversity Project. Detailed information on the distribution of all restricted-range bird species was collated. The results of the project can therefore be summarized at the global and regional levels, and also used at the local level in the planning and development of conservation projects managed by members of BirdLife International's global network.

Birds are used as indicators because they are one of the most extensively studied of the major animal and plant groups. Most of the world's bird species are already described. New bird species are currently being discovered at a rate of only 2.4 per year (Vuilleumier *et al.*, 1992) while 12.6 new mammal species have been described per year since 1982 (Wilson and Reeder, 1993). The study of birds provides a truly global perspective since they have dispersed to and diversified in virtually all regions of the world and all terrestrial habitat types and altitudinal zones. They are generally the easiest of the major animal groups to record in the field and they lend themselves particularly well to the study of species distribution and biogeography. Their taxonomy is sufficiently stable and well documented to provide a meaningful measure of diversity. The most advanced list of all 9700 of the world's bird species (Sibley and Monroe, 1990) provides a summary of the distribution and habitat requirements of almost all of these species based upon a review of the extensive ornithological literature.

BirdLife International is the organization that compiles the global Red Data Books for birds (King, 1978–9; Collar and Stuart, 1985; Collar and Andrew, 1988; Collar *et al.*, 1992). The aim of these books is to identify threatened species and collate and summarize all relevant information on their distribution and conservation status. The Red Data Book program has built up an extensive network of contacts, principally of people who are active in the field and are able to provide recent, unpublished records of threatened and restricted-range birds.

Many other sources of data are used. The ornithological literature is rich in distributional records and in information about the occurrence of species in relation to habitat. The results of ornithological expeditions and surveys are a major source of data. The localities where species have been recorded are often listed or summarized in regional works, or in studies of a particular species or group. The labels of museum specimens are another important source of data, as they often contain unpublished localities or ecological information. However, there are wide variations in the availability of data on the status of birds in different countries and regions of the world. In some countries, mainly developed countries in temperate regions, the ranges of all bird species are well understood and published in atlases which summarize the results of fieldwork by large numbers of observers (e.g. Sharrock, 1976; Root, 1988). However, in many countries there are few resident ornithologists and

relatively little recent information. Most tropical countries where the majority of restricted-range and threatened birds are found are in the latter category.

In relatively data-poor regions, the ranges of birds are inferred from all available species records together with knowledge about the habitat requirements and altitudinal range of each species. A high proportion of these records tend to be from a small number of accessible and well-known localities. It is usually necessary to interpolate and extrapolate distributions into the areas which have not been ornithologically explored. The species range maps plotted using this approach will often be accurate, but errors are inevitable if the habitat and altitudinal requirements of a species are not thoroughly understood or if an inaccurate habitat map is used.

9.3 The methodology used to identify endemic bird areas

The initial stage of the Biodiversity Project was the compilation of a list of the restricted-range bird species of the world, using the many published range maps and textual descriptions of species distributions (i.e. regional handbooks, field guides, atlases, etc.). The species taxonomy was standardized according to Sibley and Monroe (1990). Care was taken to allow for variations in quality of the available literature. For example, detailed up-to-date range maps are available for many South American species in Hilty and Brown (1986), Ridgely and Tudor (1989) and Fjeldså and Krabbe (1990), enabling an accurate candidate list to be developed for that region. In contrast, no comparable published range maps are available for continental Asia. Therefore the initial candidate list for Asia included a higher proportion of species that were later excluded because research showed that their ranges were larger than 50 000 km^2.

Comprehensive information on the distribu-

TABLE 9.1 A brief outline of the species database file structure used in this project

Field name	Description
Species	Scientific name of the species
Synonyms	A list of alternative scientific names used for this species
English	English name of the species
Family code	The family number from Morony *et al.* (1975), to enable indexing in taxonomic order
Countries	ISO codes for all countries where the species breeds
Habitat	Descriptions of breeding habitats, with references to the sources of data
Altitude	Altitudinal range during the breeding season, with references to the sources of data
Threat	Threatened or near-threatened classification (Collar and Andrew, 1988), including relevant amendments (Collar *et al.*, 1992)
Taxonomy	The few cases where the taxonomy used differs from Sibley and Monroe (1990)
EBA	Codes of the Endemic Bird Areas (EBAs) where the species breeds

tion of all of the world's restricted-range bird species was compiled and stored in computerized databases. The majority of these species occur in regions with relatively scarce ornithological data, so the databases were designed to include complete details of species distribution and ecology records. This information was analyzed to identify the Endemic Bird Areas that support concentrations of restricted-range birds. Maps were generated for the distribution of each individual species, based upon known localities and an assessment of the extent of suitable habitats. In the future, it will therefore be possible to amend a species' range map as new habitat or distributional information becomes available. If the original interpretation of the species' range was stored, rather than the data upon which it was based, it will not be possible to generate a new map in this way.

9.3.1 Database file structures

In the Biodiversity Project, compromises were made in the design of the file structure, to avoid the need for complex linkages between files and the consequent requirement for a sophisticated user interface. The data were stored in two main files, one for information relating to *Species* (Table 9.1) and the other for *Records* (Table 9.2) of species at localities.

The *Records* file was designed to include full details of all available distributional records for all restricted-range bird species. The most effective way to represent these on maps was as points since almost all records referred to an individual locality rather than to a continuous distribution (e.g. throughout a forest or along a river).

The approach used here, based upon the representation of recording localities as points, provides some useful capabilities. Points can be used at any scale. This was a major advantage in this project which aimed to present the results at the global and regional scales, and then use the same data and analyses for project development at the local scale. Any inaccuracies in the allocation of geographical coordinates are taken into account using 'certainty' and 'accuracy' codings (Table 9.2). In this approach, point data can be easily be summarized within each grid square for grid-based analysis or presentation. These data can also be analyzed in relation to polygon coverages using a GIS.

The geographical coordinates of the localities stored in the *Records* file were identified from gazetteers and using a world coverage map series at the 1:1 000 000 scale – the Operational Navigational Charts (Defense Mapping Agency, 1984–8). Once the records were georeferenced,

TABLE 9.2 The database file structure created for the distributional records

Field name	Description
Species	Scientific name of the species
Reference	Source reference for the record
Locality	Name and description of the recording locality
Country	ISO code for the recording country
Coordinates[a]	Geographical coordinates of the recording locality
Certainty[b]	Code to identify the certainty that the coordinates are for the correct locality
Accuracy[c]	Code for the accuracy of the coordinates
Record year	Year(s) of records at this locality, with a qualifier (e.g. circa) if required
Altitude	Minimum and maximum altitudes of records at this locality
Record type	Code for the type of record (e.g. collected specimen, sight record, etc.)
Status	Information on the breeding status at this locality (e.g. month(s) of records, etc.).
Notes	Information on abundance at this locality (e.g. names of observers, etc.)

[a] Represented in the database by six fields (latitude and longitude degrees, minutes, and bearing), or two fields for decimal coordinates.
[b] Recorded using the following codes: A, certain, there is an exact match for the locality name in the Gazetteer, and no possibility of ambiguity; B, probably, there is a close match for the locality in the Gazetteer, and/or ambiguity is possible, but thought unlikely; C, possible, there is a reasonable match in the Gazetteer, but a worrying ambiguity; D, unreliable, there is only a poor match in the Gazetteer, and/or unresolved ambiguity.
[c] Recorded using the following codes: A, coordinates believed to be accurate to within 5 km; B, coordinates believed to be accurate to within 20 km; C, coordinates not definitely known to be within 20 km of the recording locality.

it was possible to transfer them into a GIS to produce species maps (Figure 9.1). When suitable digital maps are available, it will be possible to further refine the boundaries of a species' distribution using elevation contours and habitat boundaries.

9.3.2 Future improvements to file structures

BirdLife International is developing a 'World Bird Database' to manage information on the conservation status of the birds of the world. The data compiled for the Biodiversity Project, the global Red Data Books, and other past and ongoing projects will form the basis of this database. A number of improvements will be made in the management of distributional data that will be based upon the experience gained in this Biodiversity Project.

One problem is data redundancy in the *Records* file, which occurs because the details of localities may be repeated in many records. This could be avoided by subdividing the

Records file into two segments: (1) a *Locality* file to include information specific to each recording locality; and (2) a simplified *Records* file, with codes to link the *Species* and *Locality* files. However, there is considerable potential for confusion when geographical coordinates are assigned to the localities where birds have been recorded. For example, a single locality may be described or spelt in many different ways or several localities within a single geographical area can have the same name. The use of the locality description alone is therefore not practical as a unique identifier in the *Locality* file.

A solution to this problem is provided by a system used in the ornithological gazetteer series that covers South American countries (e.g. Paynter and Traylor, 1977; Paynter, 1988). A recording locality is uniquely identified by its name together with the bibliographic reference in which this name was used. For example, the Santa Elena in Colombia where Nicéforo María (1947) collected is different to

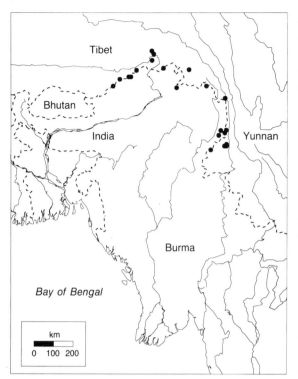

FIGURE 9.1 The distribution of Sclater's Monal *Lophophorus sclateri*. This species of pheasant is endemic to the eastern Himalayas, in northeast India, north Burma, southeast Tibet and west Yunnan province, China. It is found in rhododendron *Rhododendron* sp. and silver fir *Abies* sp. forest at altitudes between 3000 and 4000 m.

the Santa Elena where Chapman (1917) collected (Paynter and Traylor, 1981). This system has a number of advantages: it is virtually unknown for two or more localities to be described in exactly the same way within a single reference, and this prevents any possibility of ambiguity and the consequent allocation of incorrect coordinates. The description of a locality is stored in the database exactly as it occurs in the source reference, with no modification by the user. The accuracy of the coordinates allocated to a locality can be coded separately for each reference. For example, one collector may be known to have covered a wide area around a

collecting station and some birds may have come from more than 20 km away, whereas another collector may be known never to have traveled more than 5 km from the same collecting station. A disadvantage of this system is that there will be many more records in the *Locality* file, and therefore some data redundancy will be produced. There will also be a requirement to employ an additional database as a link file to cross-refer the records in the *Locality* file that come from the same places.

In the current Biodiversity Project Records file, the fields (i.e. area description, country) that hold locality descriptions are not sophisticated enough to manage this information adequately. Subdivision of the area description field into the five new fields proposed in Table 9.3 will enable more efficient storage and manipulation of these locality descriptions. For example, it will be possible to index and cross-refer records by county, province or locality name.

9.4 Project results

Restricted-range bird species comprise a significant proportion of the world's avifauna; approximately 2700 species (including 64 that have become extinct since 1800) are estimated to have ranges below the 50 000 km^2 range size threshold (i.e. almost 28% of all the world's birds). Species with small ranges are particularly vulnerable to habitat destruction and other threats; approximately 77% of the globally threatened bird species listed by Collar and Andrew (1988) have restricted ranges and were covered in the project. A total of approximately 51 000 distributional records were collated for the 2700 restricted-range species over the course of the project.

The main aim of the Biodiversity Project was to identify the places where restricted-range birds are concentrated (EBAs). The analytical techniques used to identify these areas are described in ICBP (1992). An EBA is defined as

TABLE 9.3 The database fields created to manage geographic locality descriptors

Field name	Description
Locality name	The name of the locality, as given in the reference (e.g. Emei Shan)
Qualifier	Q qualifier to the locality name that has the potential to change the coordinates allocated to the record, or to change the accuracy coding of these coordinates (e.g. '10 km south of', 'near', 'southern slopes of', etc.)
Descriptor	Additional detail provided for the locality, which may help to ensure that the coordinates are correctly allocated (e.g. '150 km southwest of Chengdu')
County	The lowest level of the national political subdivision
Province	The highest level of the national political subdivision
Country	The ISO code of the country

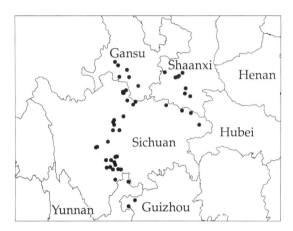

FIGURE 9.2 The central Sichuan Mountains endemic bird area. All records of the 10 restricted-range bird species of this EBA are plotted. They are found in temperate zone coniferous and mixed broadleaved and coniferous forests at altitudes between approximately 1800 and 3600 m. The distribution and habitat requirements of this group of birds closely matches those of the giant panda *Ailuropoda melanoleuca* as described by MacKinnon *et al.* (1989).

an area with at least two restricted-range species confined entirely to it. The geographical location and extent of an EBA is represented by plotting all records of the restricted-range birds within the area (Figure 9.2). A total of 221 such areas are identified world wide. Over half of these areas support less than 10 restricted-range species, but a few have over 50.

The majority of EBAs are in the tropical regions of the world and many are in tropical forests. Approximately half of the EBAs (and of restricted-range bird species) are found on islands and half are found on continental areas. The countries with the most EBAs and restricted-range birds tend to be large and in the tropics. They include the island nations of Indonesia, Papua New Guinea, the Philippines and the Solomon Islands, Peru, Colombia, Ecuador, Venezuela, Brazil, and Mexico in the neotropics.

The concentrations of restricted-range endemic birds found in EBAs are clearly of great biogeographical significance. The origins of these areas of endemism are disputed, but this debate is not a direct concern of the Bio-diversity Project, since this project aims simply to assess the conservation significance of these areas. However, it is reasonable to predict that the evolutionary circumstances which produced concentrations of restricted-range bird species will also have produced important concentrations of endemic taxa from other groups of animals and plants in the same locations. A review of the literature on endemism in other animal and plant groups found many examples of such endemic congruence. However, the documentation available on these other groups is still very incomplete (ICBP, 1992; Thirgood and Heath, 1992).

9.5 Conservation applications

EBAs are places where the destruction of relatively small areas of natural habitat could lead to mass extinctions of species. The biological importance and degree of threat to these areas are assessed in ICBP (1992). Some of these areas remain in an almost pristine state, such as the tepuís of southern Venezuela, northern Brazil, and southwest Guyana (Wege, 1989). In others, the habitats of the restricted-range birds are almost completely destroyed, such as in the Annamese lowlands of Vietnam and adjacent parts of Laos (Eames *et al.*, 1992). There is a need to investigate the state of the habitats in all of the more critically threatened EBAs using up-to-date information sources (e.g. satellite imagery).

In Colombia and Ecuador, Terborgh and Winter (1983) found little correspondence between areas with concentrations of restricted-range birds and existing or proposed protected areas. ICBP (1992) found that the coverage afforded to EBAs by the global network of protected areas varies widely. Some of these areas have meager official protection, such as the Marañón valley EBA in northern Peru. A few are almost completely protected, such as the Galápagos Islands EBA in Ecuador (IUCN, 1992). The majority of the EBAs have a limited area under official protection that is generally less than 10% of their total areas, while the mean figure for all EBAs is 8%. The detailed distributional information compiled during the course of this project will enable the identification of many important unprotected areas. A high priority for BirdLife International's future research, advocacy, and program work will be to improve protected area networks and to promote the protection of all EBAs.

The EBAs on continents and large islands usually represent a discrete zone of habitats that are relatively isolated from other similar habitats. The use of the 50 000 km^2 range size criterion has led to the identification of areas of a limited size as EBAs. These are places where relatively limited habitat destruction or modification could lead to species extinctions. The results of this project provide a blueprint

for the conservation of many of the most threatened components of the world's biodiversity.

Many bird species with range sizes above 50 000 km^2 have strict habitat requirements and are restricted to a particular vegetation zone. The application of this methodology to these more widespread species can provide insights into the distribution patterns of these species relative to vegetation types. Such an application could be used to map the species which are characteristic of a particular vegetation zone, in order to produce a detailed map of the zone itself. This could be used to assess habitat destruction and adequate representation in the protected areas network.

9.6 An effective use for biodiversity data

During the course of the BirdLife Biodiversity Project, a large database was developed with information about restricted-range bird species and their distributions. This database was to identify and document Endemic Bird Areas (EBAs). It also has many other potentially useful conservation applications. At the time that this project started there were major limitations in the use of GIS software and the associated computer hardware. Since then, rapid improvements in the performance of this software and faster, cheaper processors and memory have become available. The availability of georeferenced, high-quality datasets will increasingly become the limiting factor in GIS technology.

The ornithological literature, museum bird collections and the expertise of individual ornithologists represent a more complete database than is available for any other major faunal group. This project is an example of how these information sources can be tapped to compile a data set with a wide array of conservation applications.

9.7 Summary

The BirdLife* Biodiversity Project has mapped the distributions of all of the world's restricted-range bird species, defined as those with ranges estimated to total less than 50 000 km², which comprise almost 28% of all bird species. The majority of these bird species occur in relatively poorly known tropical regions of the world, and databases have been designed to store full details of all available distributional records of these species and their habitat requirements and altitudinal ranges. The information compiled has been used to identify 221 Endemic Bird Areas (EBAs). A literature review found evidence that many of these areas also support important concentrations of endemic taxa from other groups of animals and plants. EBAs are therefore key areas for biodiversity conservation. They can be used to help assess the effects of habitat destruction and modification on global biodiversity, to evaluate existing protected areas systems, and to identify areas for designation as new protected areas. The methodology developed in this project can also be applied to more widespread species, to investigate the patterns in their distributions, to help assess their conservation status.

Acknowledgements

I am grateful to Colin Bibby, Nigel Collar, Alison Stattersfield and Adrian Long at BirdLife International for their comments on earlier drafts of this chapter. Adrian Long also helped in the preparation of the maps, and he and Henk van Dijkhuizen, Carsten Rahbek and Jon Fjeldså of the Zoological Museum of Copenhagen took part in valuable discussions on database file structures.

References

Chapman, F.M. (1917) The distribution of bird-life in Colombia; a contribution to a biological survey of South America. *Bulletin of the Amererican Museum of Natural History*, **36**.

Collar, N.J. and Andrew, P. (1988) *Birds to watch: the ICBP world checklist of threatened birds*, International Council for Bird Preservation (Techn. Publ. 8), Cambridge, UK.

Collar, N.J. and Stuart, S.N. (1985) *Threatened Birds of Africa and Related Islands: The ICBP/IUCN Red Data Book* (3rd edn), Part 1, International Council for Bird Preservation/International Union for Conservation of Nature and Natural Resources, Cambridge, UK.

Collar, N.J., Gonzaga, L.P., Krabbe, N. *et al.* (1992) *Threatened Birds of the Americas*: *The ICBP/IUCN Red Data Book* (3rd edn), Part 2, International Council for Bird Preservation, Cambridge, UK.

Defense Mapping Agency (1984–8) *Operational Navigation Charts*, Defense Mapping Agency Aerospace Center, St. Louis, Miss.

Eames, J.C., Robson, C.R., Nguyen Cu and Truong Van La (1992) *Vietnam Forest Project: Forest Bird Surveys in Vietnam*, International Council for Bird Preservation (Study Report 51), Cambridge, UK.

Fjeldså, J. and Krabbe, N. (1990) *Birds of the High Andes*, University of Copenhagen Zoological Museum, Copenhagen.

Hall, B.P. and Moreau, R.E. (1962) A study of the rare birds of Africa. *Bulletin of the British Museum of Natural History*, **8**, 313–78.

Hilty, S.L. and Brown, W.L. (1986) *A Guide to the Birds of Colombia*, Princeton University Press, Princeton.

ICBP (1992) *Putting Biodiversity on the Map: Priority Areas for Global Conservation*, International Council for Bird Preservation, Cambridge, UK.

IUCN (1992) *Protected Areas of the World: A Review of National Systems*. Vol. 4: *Nearctic and Neotropical*, International Union for Conservation of Nature and Natural Resources, Gland, Switzerland and Cambridge, UK.

King, W.B. (1978–9) *Red Data Book*, 2. *Aves* (2nd edn), International Union for the Conservation of Nature and Natural Resources, Morges, Switzerland.

MacKinnon, J., Bi Fengzhou, Qiu Mingjiang *et al.* *National Conservation Management Plan for the Giant Panda and its Habitat*, Ministry of Forestry and WWF – World Wide Fund for Nature, Beijing, China and Gland, Switzerland.

McNeely, J.A., Miller, K.R., Reid, W.V. *et al.* (1988) *Conserving the World's Biological Diversity*, IUCN, WRI, CI, WWF-US, The World Bank, Gland and Washington.

Mittermeier, R.A. (1988) Primate diversity and the

* As of March 1993, the International Council for Bird Preservation (ICBP) has operated under the new name of BirdLife International.

tropical forest: case studies from Brazil and Madagascar and the importance of the megadiversity countries, in *Biodiversity* (ed. E.O. Wilson), National Academy Press, Washington.

Morony, J.J., Bock, W.J. and Farrand, J. (1975) *Reference List of Birds of the World*, American Museum of Natural History, New York.

Myers, N. (1979) *The Sinking Ark*, Pergamon Press, Oxford.

Myers, N. (1988) Threatened biotas: 'hotspots' in tropical forests. *Environmentalist*, 8, 187–208.

Myers, N. (1990) The biodiversity challenge: expanded hot-spots analysis. *Environmentalist*, 101, 243–56.

May, R.M. (1990) How many species? *Philosophical Transactions of the Royal Society London* 330, 293–304.

Nicéforo María, H. (1947) Notas sobre aves de Colombia, II. *Caldasia*, 4, 317–77.

Paynter, R.A. (1988) *Ornithological Gazetteer of Chile*, Museum of Comparative Zoology, Cambridge, Mass.

Paynter, R.A. and Traylor, M.A. (1977) *Ornithological Gazetteer of Ecuador*, Musuem of Comparative Zoology, Cambridge, Mass.

Paynter, R.A. and Traylor, M.A. (1981) *Ornithological Gazetteer of Colombia*. Museum of Comparative Zoology, Cambridge, Mass.

Ridgely, R.S. and Tudor, G. (1989) *The Birds of South America: The Oscine Passerines*, Vol. 1, Oxford University Press, Oxford and Tokyo.

Root, T. (1988) *Atlas of Wintering North American Birds: An Analysis of Christmas Bird Count Data*, Univ. of Chicago Press, Chicago and London.

Sharrock, J.T.R. (1976) *The Atlas of Breeding Birds in Britain and Ireland*, Poyser Ltd, Calton, Staffordshire, UK.

Sibley, C.G. and Monroe, B.L. (1990) *Distribution and Taxonomy of Birds of the World*, Yale University Press, New Haven and London.

Stattersfield, A.J., Crosby, M.J., Long, A.J. and Wege, D.C. (in prep.) *A Global Directory of Endemic Bird Areas*, BirdLife International, Cambridge, UK.

Terborgh, J. and Winter, B. (1983) A method for siting parks and reserves with special reference to Colombia and Ecuador. *Biological Conservation*, 27, 45–58.

Thirgood, S.T. and Heath, M.F. (1992) Global patterns of endemism and the conservation of biodiversity, in *Systematics and Conservation Evaluation* (ed. P.J. Forey, C.J. Humphries and R.I. Vane-Wright), Oxford University Press, UK, pp.207–27.

Vuilleumier, F., LeCroy, M. and Mayr, E. (1992) New species of birds described from 1981 to 1990. *Bulletin of the British Ornithology Club Centenary Suppl.*, 112A, 267–310.

Wege, D.C. (1989) Conserving biological diversity: identifying areas of avian endemism. A case study from the tepuis region of southern Venezuela and contiguous countries. Unpublished M.Sc. thesis.

Wilson, D.E. and Reeder, D. (1993) *Mammal Species of the World: A Taxonomic and Geographic Reference*, Smithsonian Inst., New York.

Wilson, E.O. (1988) The current state of biological diversity, in *Biodiversity* (ed. E.O. Wilson), National Academy Press, Washington, D.C., pp.3–18.

Chapter ten

Mapping and GIS analysis of the global distribution of coral reef fishes on an equal-area grid

Don E. McAllister,
Frederick W. Schueler,
Callum M. Roberts and
Julie P. Hawkins

10.1 Introduction

10.1.1 Review of the diversity and status of coral reefs and fishes

Tropical coral reefs form the most diverse and productive ecosystems in the oceans (Dubinsky, 1990). At higher taxonomic levels (e.g. orders, classes and phyla) reefs are perhaps the most diverse ecosystems in the world and they support a vast, morphologically diverse and colorful array of species. An estimated 4000 species of fishes, approximately 25% of the marine fish species, inhabit coral reefs (McAllister, 1991a). Despite this high biodiversity, coral reefs cover only 0.18% of the world's oceans (Smith, 1978). The fish fauna of coral reefs is thus two orders of magnitude richer than the average of the fish diversity in the oceans. Coral reefs are important to people

Mapping the Diversity of Nature. Edited by Ronald I. Miller.
Published in 1994 by Chapman & Hall, London. ISBN 0 412 45510 2.

since they provide more than half the animal protein in the diets of many tropical countries, considerable employment, and coastal protection against storm waves (McAllister, 1988).

If we are to effectively protect marine species along with the ecological benefits provided by these organisms, data about the geographic distribution of biodiversity hotspots is required to evaluate this most important resource. Such an evaluation is especially critical now, in order to best utilize the limited funds available from governments, NGOs, and international agencies for marine conservation. Current identification of megadiversity countries (McNeely *et al.*, 1990) is largely based on terrestrial organisms (e.g. mammals, birds, reptiles, amphibians, butterflies and angiosperm plants). Increases in numbers of species toward the equator are firmly established in terrestrial ecosystems, but Clarke (1992) concluded that evidence for a latitudinal cline is lacking for the majority of marine taxa.

Coral reefs are found between 32°N and 32°S latitude in ocean waters of 20°C or warmer. The majority of reefs are shallower than 30 m depth and they border the coastlines of continents, islands and archipelagos, mostly within the waters of developing nations (UNEP/IUCN, 1988). This distribution, close to large, growing human populations (Harrison, 1992: 198), subjects many coral reefs to intense human impacts.

Coral reefs have formed and persisted only where they are able to recover from natural episodes of physical and biological stress phenomena such as storms, freshwater, cold and outbreaks of predators. Reefs have not formed or have died where natural stresses are chronic. Some anthropogenic stresses (e.g. oil spills and nuclear explosions) are episodic. Others (e.g. sedimentation and sewage) are typically chronic and they may co-occur with other stressors. Recovery under chronic stresses is usually unlikely. As a consequence, 70% or more of the reefs in Japan (Planck *et al.*, 1988),

the Philippines (Gomez and Alcala, 1979), and in Costa Rica (Robinson, 1987) are in only poor or fair condition by the criteria of Gomez and Alcala (1979).

Anthropogenic stresses on coral reefs are summarized by UNEP/IUCN (1988), Smith and Buddemeier (1992), Wells and Hanna (1992) and Weber (1993). Sedimentation from inappropriate agricultural, development and forest management practices is one of the leading stressors in many countries (McManus, 1988; Planck *et al.*, 1988; Samoilys, 1988). Destructive fishing techniques, including explosives, fish poisons and trawls, are degrading coral reefs in Southeast Asia, East Africa and the Caribbean (Rubec, 1986; Alcala and Gomez, 1987; Ansula and McAllister, 1992). Coral mining for landfill, building materials, and for the manufacture of lime removes the substance of reefs in many countries. Many other stressors are summarized in the cited reviews. Weber (1993) maps the degrees of threat to global reef areas.

It is thought that widespread coral bleaching (i.e. loss of colored symbiotic alga or their pigment from the coral polyps) in the Indo-Pacific and Caribbean reefs during the 1980s was caused by global warming (Goreau, 1990; Glynn, 1991; Smith and Buddemeier, 1992; but see D'Elia *et al.*, 1991). Coral bleaching can result from a 1–2 day exposure to 3–4°C above the normal daily temperature maximum or exposure to several weeks of exposure to temperatures that are 1–2°C above the normal daily maximum (Smith and Buddemeier, 1992). Mortality is > 90% for temperature elevations of 4°C for even a few hours.

10.1.2 The information crisis

Compared with terrestrial ecosystems, marine conservation suffers from a lack of attention and resources (Norris, 1993), and although the decline of coral reef fishes and their habitat is apparent, documentation is scattered and uneven. The establishment of underwater marine parks and reserves has been given a lower priority than those on land (reviewed by Randall, 1982), in part because of human bias in favor of land and in part because of greater human impacts on land. No coral reef fish species is classified as threatened or endangered throughout its range, principally, in our view, because scant attention has been devoted to the question (the jewfish, *Epinephelus itajara*, has been declared endangered in the US Virgin Islands). Declining numbers of large individuals and population sizes of certain species have been observed and documented in some areas, and of ca. 100 countries with coral reefs, 80 have reported overfishing (Wells and Hanna, 1992).

Economic overfishing is usually distinct from the danger of species-wide extinction. Indeed, some authorities (Randall, 1980) deny that coral reef fishes could become endangered, due to their wide geographic ranges (many species range from East Africa to the mid-Pacific, and some are pan-tropical). Even widespread and common species may become endangered or extinct if human impacts are widespread and chronic, and the total area of coral reef occupied within a range spanning several thousand kilometers may be quite small. This is especially true in archipelagos such as Micronesia and Polynesia. Furthermore, population densities of many species are low. Thus at Toliara, southwest Madagascar, only 25% (136) of the species were ranked as abundant (Groombridge, 1992). This pattern of species abundance is seen in most speciose communities (Magurran, 1991).

Until as little as 10 years ago, insufficient information was available for large-scale mapping of the distributions of coral reef fishes. There were few regional ichthyofaunal handbooks for identification and many groups were without monographic taxonomic treatments. Publication of increasing numbers of handbooks, many with photographs of live fishes,

and of revisions of genera and families, has produced an improvement in the accuracy of species lists published for reef areas. However, the taxonomic description of the world's ichthyofauna is incomplete and each year more than 100 new species, several genera, and the occasional family are described. Many of these newly described taxa are from coral reefs. Summaries of threatened coral reef species or of areas with high biodiversity and endemism are not currently available. There is only an estimate of the number of species of fishes regularly inhabiting coral reefs (McAllister, 1991a) and there is no published world list of fishes or coral reef fishes (DEM has a working manuscript list of fishes of the world). Environmental data, when available, are seldom assembled in ways that facilitate comparisons with biotic data. These problems are compounded by current reduced support for biosystematic research in the north, and the lack of funding for natural history museums and biosystematic research in the south.

10.1.3 The global perspective

All of the above concerns led the Species Survival Commission of IUCN to establish a Coral Reef Fish Specialist Group charged with promoting the conservation of coral reef fishes, obtaining sound information on the status of these fishes, evaluating human impacts and making recommendations for their conservation. Ocean Voice International, a non-profit marine environmental organization, was chosen to manage this project for the Coral Reef Fish Specialist Group. Challenged by the scale of the problem and a lack of information on coral fish biodiversity patterns, we developed and are enhancing a GIS that is similar to the GIS developed for the recent ICBP biodiversity mapping study (ICBP, 1992, ch. X).

Fish, like land birds, are better known taxonomically than invertebrates, non-vascular plants, and microorganisms; they can therefore act as a provisional 'flagship' group for coral reefs. The coral reef fish GIS which we describe in this chapter is designed to provide information on where the biodiversity and threats to diversity are located, and so to focus efforts on fish habitat protection where it is most needed. We believe that fish are only a provisional flagship group, because coral reefs are such distinctive and important habitats that biosystematic studies, biological inventories, and biogeographic analyses should be carried out to evaluate the status of every major taxon.

Biogeographic hotspots are areas which support large numbers of species compared to the surrounding region. These are clusterings of species that create strong regional anomalies in the global increase in biodiversity toward the equator (e.g. Clarke, 1992; McAllister *et al.*, 1986, North American freshwater fishes; Anderson, 1984a, b, North American amphibians, reptiles and birds). Hotspots may occur where several biogeographic zones abut (e.g. northwest Australia/southern Indonesia) and where several tectonic plates meet and biotas from adjacent plates come in contact or mix (Woodland, 1986; Springer, 1982; Planck *et al.*, 1988). Hotspots also occur in areas where conditions favor rapid speciation or in areas with relatively stable conditions that reduce rates of extinction.

Hotspots for different taxa often do not coincide. The principal North American hotspot for freshwater fishes is in the Cumberland–Tennessee Plateau drainages of the eastern United States, while for birds it is in northern Mexico and the adjacent areas in the United States. This means that different protected areas may have to be chosen for the different taxa to maximize conservation of biodiversity. Establishing reserves for tigers may protect some of the less widely ranging species, but many species may be lost if these reserves do not fall in regional hotspots for the other taxa. Genera

of reef-forming (hermatypic) corals are mapped (Vernon, 1986, 1993), and the addition of maps for fish, perhaps the reef animals most unlike corals, should improve judgements about the total diversity of reefs. This will be an important step in selecting marine reserves or parks and for devising additional conservation strategies.

Fundamental changes are evolving in the philosophy of biodiversity conservation. Conventional conservation focuses on a few higher large-sized vertebrates and plants. Today a broader concept of biodiversity is becoming accepted. There are too many species at risk to classify as rare or endangered, let alone to prepare and carry out recovery programs for each species. One response to this problem is a call for the conservation of ecosystems, landscapes and seascapes, to protect numerous species and their habitats. This approach is valid, but it does not include a method to identify which ecosystem, landscape or seascape should be set aside as a protected area for optimal conservation.

Mapping and locating species hotspots offers a solution to this dilemma. But this approach ignores the phylogenetic differences between higher taxa, such as genera, families and orders (McAllister, 1991b). The species hotspot approach fails to distinguish between an area with three species of sparrows and another with a sparrow, an eagle and a swallow, despite the greater taxonomic, ecological and genetic diversity of the latter grouping. A resolution of the dilemma is simple. Compare the total number of taxa (species, genera, families, etc.) found in two prospective areas for conservation (or when cladistic analyses are available, total the number of cladogram nodes between species). This eco-taxa hotspot approach is discussed in more detail by Vane-Wright *et al.* (1991), Williams (1993), Williams and Humphries (1992), Williams and Gaston (in press) and McAllister (1993). The eco-taxa approach is relatively simple and yet it is powerful in combining information from the two disciplines of biosystematics and ecology.

Except for a few widespread and conspicuous species which can be inventoried by remote sensing or repeated sampling schemes, knowledge of biotic distribution depends on analysis of individual records. A record, traditionally a museum specimen, is a documented verifiable encounter with a species at a particular place and time. Spot distribution maps represent each record with a symbol instead of using shading or encircling a presumed known range. Although such maps are available in some monographs for coral fishes (e.g. Allen, 1975), there are no published maps for most species.

We began our work by preparing spot distribution maps of >950 coral reef fishes, about a quarter of the known species (Roberts and Hawkins, 1992). Beginning with maps compiled by the late Sir Peter Scott, one of the founders of the IUCN's Species Survival Commission, records arc assembled from taxonomic and regional publications and from museum specimens and they are plotted onto the equal-area basemaps described by Alfonso (1991).

The fish distribution maps show geographic patterns for individual species. McAllister *et al.* (1986) earlier used computers and software to assemble fish distributional and environmental data sets into grids and to carry out analyses within and among these data sets. This approach is now automated by the GIS and packages of programs, ranging in cost from a few hundred to many thousands of dollars, are now available commercially to handle data and cartographic manipulations. We use a GIS package (QUIKMap running in an inFocus module) in this study, but one of the several advantages of the equal-area grid GIS approach described here is that moderate-sized data sets can be compiled and manipulated manually.

10.1.4 GIS tools and conservation

(a) GIS parameters

Geographic units (e.g. countries) vary in size and shape and seldom coincide with trends or variables that one wishes to investigate. In addition, different sets of point data seldom coincide. The grid system is found to be advantageous in stabilizing the unevenness of sampling effort (McAllister *et al.*, 1986) when trend surface analyses (from Schueler, 1982) are used to estimate environmental parameters between known data points. The establishment of a standard global grid would enable researchers to more easily exchange data sets, reduce duplication and labor, and facilitate comparison of results.

Quantitative biogeographic analysis, like all mapping, has always been hampered by the sphericity of the earth. McAllister *et al.* (1986) uses cells one degree of latitude by one degree of longitude that decrease in size and 'squareness' toward the pole. The usual method to produce a grid of 'square' cells for a local arca (i.e. the Universal Transverse Mercator Grid) is based on a system of narrow longitudinal zones, each with its own grid system (Sebert, 1975; Weller and Oldham, 1988). This system, however, is awkward for global comparisons.

(b) Equal-area cell system

To create a global system of essentially square, equal-area cells, we sacrifice the requirement that the longitudinal bounds of the cells line up to produce a Cartesian grid. Therefore the system that we use to spatially define the cells does not provide a coordinate system for computations. Consequently we use the latitude and longitude at the center of each cell to locate it in spherical geometry or on a two-dimensional map projection. This equal-area cell system is based on the Sanson–Flamsteed map projection. In this projection longitude is collapsed onto a central meridian (i.e. east–

west distance from the meridian = longitude × cosine (longitude)) so that relative area is preserved and the parallels of latitude are straight lines perpendicular to the central meridian.

Research hypotheses and data impose restrictions on cell size for any project. A fine grid may be required to observe effects at a small geographic scale, but may not give better resolution than a coarser grid if sampling has been relatively sparse or uneven (McAllister *et al.*, 1986). We selected the 2-degree cell system as the best compromise for obtaining locally useful results and for adequately spacing the records. A 2-degree cell has an area of about 50 000 km^2. The cell bands are bounded by even-numbered degrees of latitude and originate at the 0° meridian of longitude. Systems of cells are propagated east and west from this central meridian, and must meet somewhere, not necessarily at 180° (the last cell is usually smaller). We place this overlap zone in areas of minimal biogeographic and political interest (i.e. the tropical eastern Pacific, where there are no islands, and then west running northward through the Bering Strait). The overlap zone could be relocated for another study.

(c) GIS software and hardware used

We examined different moderately priced (under US$1500) GIS packages for desktop microcomputers. This appraisal, based upon capability needs and user-friendliness, selected Earth and Ocean's inFOcus (Earth and Ocean Research Limited, 1991) that runs QUIKMap by Ayxs software (Environmental Sciences Limited, 1990) with FoxPro programs (Fox Software Inc., 1991). This GIS will run on a microcomputer with 525 kilobytes (kb) of random access memory (RAM), a VGA-monitor, and a 50 megabyte (Mb) hard disk. Memory above the DOS 640 kb limit will be utilized, if it is configured for EMS-memory. We ran it initially on a 386 20 MHz central processing unit (CPU) system with 8 Mb RAM,

a 100 Mb hard disk and a VGA-monitor. We are now running it at 66 MHz internally and 33 MHz externally on an Intel 486 CPU with internal math coprocessor, a memory cache, 4 Mb RAM, a 200 Mb hard disk and a SVGA-monitor. The new system draws maps on the screen faster, saving time for the operator. The map can be saved for re-use, revision, zooming in on details, and printing or plotting. Maps can be labeled, printed or displayed on a variety of black or color printers and plotters.

Earth and Ocean prepared, on contract for Ocean Voice, a module incorporating Ocean Voice's equal-area grid. The world basemap provided by Earth and Ocean shows the location of most of the coral reefs of the world as a discrete colored map layer. The GIS can import other basemaps, switch between several map projections, zoom in on localities to provide more detail, and output on several printers and plotters. One can key information into the database, digitize localities from a map, create or edit records using a cursor on a map displayed on the computer screen, and transfer data between compatible databases. Although our initial goal was to study coral reef ecosystems in the tropics, the equal-area grid works equally well for most terrestrial areas.

The data are stored as latitude–longitude points in a generic FoxPro '.DBF' file and this allows the point locations to be used for other projects. Any data with either these or UTM coordinates can be imported into the system. Manipulation of the database then enables the number of species in a cell to be used for various analyses. This includes summation of any combination of species or families, counts of endemics and output of appropriate maps. These manipulations are done with FoxPro routines outside the inFOcus mapping module.

Table 10.1 lists the longitudinal width of 2-degree cells for a range of latitudes. Mean latitudes in this table are lower than median cell band latitudes (odd-numbered degrees)

TABLE 10.1 A comparison of areas comprising 2-degree cells. These are calculated for a spherical model of the globe and then corrected for an oblate spheroid. Mean latitude is the latitude of the centroids of each cell band. Corrected longitudinal width is the width of cells adjacent to the Greenwich meridian, as calculated by Earth and Ocean on inFOcus. The relative area is the ratio corrected area/spherical area

Latitude		Longitude		
Cell band	Mean	Corrected area	Spherical area	Relative area
1	1.00000	1.00015	1.00000	1.00015
11	10.99865	1.01859	1.01856	1.00003
21	20.99074	1.07962	1.07098	0.99966
31	30.97112	1.16527	1.16646	0.99898
41	40.93607	1.32189	1.32481	0.99780
51	50.88377	1.58202	1.58877	0.99575
61	60.81444	2.04589	2.06235	0.99202
71	70.73032	3.02238	3.07109	0.98414
81	80.63541	6.13179	6.39148	0.95937

because cells bounded by meridians are wider toward the equator. The number of cells in a cell band ranges from 180 cells at the equator to 136 at 41° latitude and 30 at 81° latitude. Earth and Ocean calculates a system of 2-degree cells designed to have equal areas on the 1927 North American Datum spheroid (a standard model of the flattened shape of the earth) used by QUIKMap. Table 10.1 compares the area of these cells to those based on a spherical earth. Up to a latitude of 61° the areas of cells agree within 1%, and even at 81° they agree within 5%. We use the corrected 2-degree cell system to obtain the results reported here but the simpler spherical model would have been sufficient.

There are about 5300 cells between 30°N and 30°S latitude but only about 700 contain coral reefs. Proportionally, about 64% of these are in the Pacific Ocean, 21% in the Indian Ocean and 15% in the Atlantic Ocean. The ratio of Indo-Pacific (85%) to Atlantic cells (15%) is close to the ratio for coral reef areas given by Newell (1971): 1 250 000 km^2 (83%) for the Indo-Pacific and 250 000 km^2 (17%) for the Caribbean. Fish data currently occupy 480 cells or about 68% of the known reef cells.

A few of the mapped fishes occur in cells having rock rather than coral reefs (e.g. Ascension and St Helena).

10.2 Methods

10.2.1 Map data collection, entry and processing

In this study about 950 species of coral reef fishes are mapped from eight families: surgeon-fishes, Acanthuridae; butterflyfishes, Chaetodontidae; neon gobies, *Gobiosoma*, Gobiidae; wrasses, Labridae; angelfishes, Pomacanthidae; seabasses, Serranidae (*sensu lato*); pufferfishes, Tetraodontidae; and moorish idols, Zanclidae. For the study reported in this chapter, 8928 individual records were entered into the database for 799 of these species, including 12 Gobiidae, 264 Serranidae, 260 Labridae, 114 Chaetodontidae, 1 Zanclidae, 63 Acanthuridae and 85 Tetraodontidae. These 799 species comprise the all-family sample that includes about 20% of all coral reef species and 7 of the 46 primarily reef-associated families (Thresher, 1992). More than 110 fish families are found

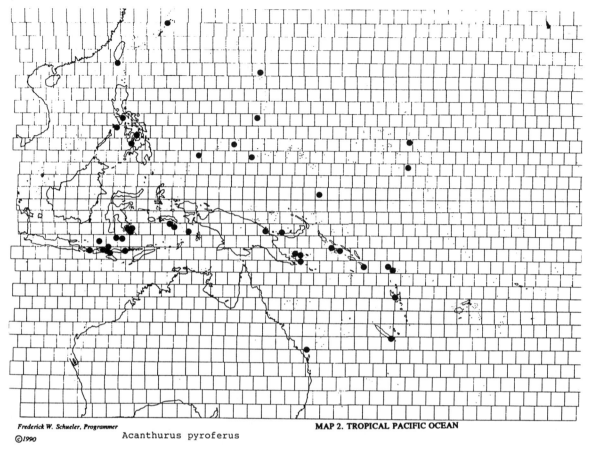

Frederick W. Schueler, Programmer
©1990   Acanthurus pyroferus   MAP 2. TROPICAL PACIFIC OCEAN

FIGURE 10.1 Pacific Ocean records of the mimic surgeonfish, *Acanthurus pyroferus*. This shows the form in which data were entered into the GIS database.

on coral reefs, but most have relatively few coral reef species. Figure 10.1 shows a sample spot distribution map for pacific records of the mimic surgeonfish *Acanthurus pyroferus*.

Distributional data for the maps are compiled from 66 sources. Sir Peter Scott's lists of species present at sites throughout the Indo-West Pacific during the 1970s and early 1980s are carefully checked using his detailed field descriptions and up-to-date taxonomic works (Roberts, 1987; Roberts and Hawkins, 1992). The map data are then transferred to the Ocean Voice equal-area grid maps, and records are added from other published and unpublished sources. A record or identity of a fish is not

used when the accuracy is in doubt. Map data for the 950 species are still being gathered, so future analyses will be based on more records.

Faunal studies are uneven and some are out of date. Recent faunal works or catalogues provide some degree of confidence for the Red Sea, southern Japan, the Bahamas, the Great Barrier Reef, and the Chagos Archipelago. On the other hand, studies in possibly important areas like Somalia, Mozambique, parts of Polynesia and the eastern tropical Pacific are either non-existent or focused upon only the commercial fishes.

A number of works (e.g. Myers, 1989) attribute records to island groups rather than specific localities. Such records link with an arbitrary point in the center of the area and similar records for each species are plotted on this point. The maps thus represent a mix of very precise records and those which are more broadly defined. In these maps, species appear to be more widespread but this is probably attributable to lower species/cell totals that represent areas with records from individual collections rather than regional summaries.

The Chaetodontidae, for example, are well studied taxonomically, while there is no taxonomic monograph of the very diverse Gobiidae, although several of the genera are revised. Some areas (Japan, Israel, Natal, Florida, etc.) have high concentrations of ichthyologists or biological stations which have encouraged visits. All these factors augment or diminish the number of species known from an area and consequently distort the analysis results. Nevertheless, previous experience (McAllister *et al.*, 1986) leads us to believe that grid-based species density maps are a robust method of locating hot spots. Subsamples of 311 and 501 of the 775 species of North American freshwater fishes and 125 endemic species all show the same major hotspots.

Map data (Figure 10.1) are entered into the computer using a mouse-driven cursor. The ability of QUIKMap to zoom in to any scale helps increase precision and about 80% of the points are entered from an enlarged basemap. For the western Pacific we enter records by stations since many of the records occupy the same locations. About 4% of the records are duplicates in the present data set.

(a) Hotspots and endemic species

Analyses are carried out for all species (Plates 7–8) and endemic species (Figure 10.3). ICBP (1992) defines endemic species as those with ranges of less than 50 000 km^2. Each of the grid cells on our maps correspond to this area. However, marine species with planktonic dispersal tend, on average, to have much larger ranges than terrestrial species and so a range size of 50 000 km^2 would be inappropriate for fishes. For the purposes of this study we define endemics as those species restricted to an area with a diameter of 4 cells or fewer (i.e. an area of 12.5 cells or fewer).

We compile the number of species per cell for each family and for all families. To obtain a regional picture of species density, we prepare uniform maps in which each cell contains records of every species found in adjacent cells (calculated as cell centers within 3° of the center of the cell being summed). This helps compensate for the clumping of records at arbitrary archipelago centers.

We do not deal with patterns of similarity among cells or regions. Our hotspots are not biogeographical provinces but contiguous areas with fairly uniform high numbers of species. Likewise, our endemics are species with small ranges, not species with ranges limited to a geographic or biogeographic area.

(b) Latitude–longitude clines

We sum the species present in the 10° longitudinal bands (Figure 10.2) to test the hypothesis that species are evenly distributed around the tropics. We sum species in 4° latitudinal bands (Table 10.2) to test the hypothesis that the number of species increases toward the equator. Product–moment correlations test the relationship between the number of species and the latitude of the band. The tests are applied to several groupings of cell bands: (1) from the equator northward, (2) from the equator southward, (3) poleward from the equator (i.e. north and south latitudinal bands together), and (4) poleward from the equator excluding the bands immediately north or immediately south of the equator.

(c) Range sizes

To test the hypothesis that most coral fishes are widespread, and hence at little risk of extinc-

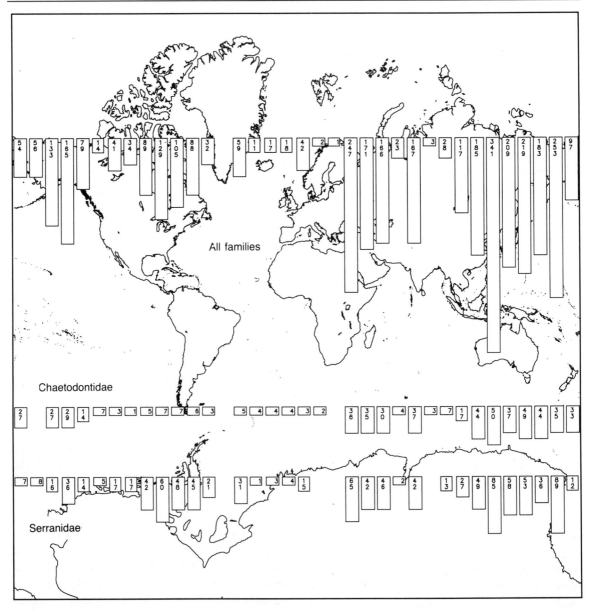

FIGURE 10.2 Histograms of the longitudinal species density (i.e. species numbers/10° longitudinal band) around the world for the butterflyfish and seabass families. The number of species is shown at the top of each bar of the histogram. Note that seabass (Serranidae) samples are relatively richer in species than butterflyfish (Chaetodontidae) samples in the western Atlantic.

tion, we measure the range span of each species as the angular distance between the centers of its most distant occupied cells. For species with ranges < 180°, this is calculated as the maximum angle (the great circle distance) between records. For species with larger ranges we use

TABLE 10.2 The occurrence of all families of coral reef fish species in latitudinal ranges across the globe

Latitude	Number of species	Latitude	Number of species
32–28°N	265	0–4°S	116
28–24°N	344	4–8°S	275
24–20°N	296	8–12°S	295
20–16°N	215	12–16°S	234
16–12°N	337	16–20°S	301
12–8°N	314	20–24°S	288
8–4°N	309	24–28°S	138
4–0°N	178	28–32°S	150

the shortest distance between records that crossed the greatest longitudinal gap in the range (Table 10.3).

(d) Differences in patterns between taxa

To test the 'tiger reserve hypothesis' that a reserve chosen for one species or taxonomic group effectively protects many others, we compare latitudinal and longitudinal trends for two family pairs. For latitude we compare the number of species of butterflyfishes, Chaetodontidae, and wrasses, Labridae, in 4° bands between 32°N and 32°S. For longitude, we compare the number of species of Chaetodontidae and seabasses, Serranidae, in 10° longitudinal bands around the globe (Figure 10.2).

(e) Higher taxon diversity

Using this same approach, we sum genera and families in each occupied cell and we calculate correlations between the number of species, genera, families and total taxa (i.e. species + genera + families).

10.2.2 Analysis

(a) Other data sets

The next stage in our project is to gather climatic, geologic, oceanographic and biologic information about the environment of coral

reef fishes and to determine the location and severity of threats to fishes and reef habitats. To preserve biodiversity, we attach the highest conservation priority to areas where either reef fish diversity and/or endemism is high and human impacts are greatest.

(b) Statistical methods

Correlation coefficients were calculated between numbers of species in each family and total species, and latitude and longitude. Some data sets were compared using chi square tests. The shapes of latitude–longitude distributions were compared by 2-sample Kolmogorov–Smirnov tests (Sokal and Rohlf, 1981). A critical probability of $p=0.05$ was used. Figure 10.2 and Table 10.2 show latitudinal and longitudinal patterns of species richness, and the global species density maps, Plates 7–9 show the locations of hotspots.

10.3 Results and discussion

10.3.1 Species density: hotspots and clines

Plates 7–9 present the species density changes obtained in our analyses. The species densities on the maps in these plates reflect changes both in the level of sampling and in the biodiversity densities. The presented analyses of diversity clines are therefore performed at a level coarser than single cells to offset this sampling intensity bias. Further data will increase the accuracy of these mapped gradients.

(a) Hotspots

Inspection of the smoothed species density maps (Plates 7–9) and the latitudinal and longitudinal cline patterns (Table 10.2, Figure 10.2) reveal the locations of some global coral reef fish hotspots. The highest cell totals are found in the 'coral triangle' (cf. Briggs, 1974: 21) with cell counts well above 100 found from

TABLE 10.3 A comparison of coral reef fish species richness in relation to the species diverse range spans. The increasing number of coral reef fish species is shown in relation to the increase in the cumulative percentage. Range spans are defined as the angular distance between the most distant occurrence records

Geographic range		Species number		
Angle (degrees)	Range span (km)	Number of species	Cumulative number	Percent
0–2°	0–222	133	133	17%
<5°	223–555	12	145	18%
<10°	556–1 110	33	178	22%
<20°	1 111–2 220	86	264	33%
<40°	2 221–4 440	110	374	47%
<60°	4 441–6 660	100	474	59%
<90°	6 661–9 990	98	572	72%
<150°	9 991–16 650	112	684	86%
<180°	16 651–19 980	11	695	87%
<240°	19 981–26 640	13	708	89%
<300°	26 641–33 300	7	715	89%
≥300°	33 301–40 000	84	799	100%

the Ryukyu Islands of southern Japan, southeastern New Caledonia, in the eastern Indo-Australian Archipelago, and in Melanesia. The Ryukyu Islands have the highest smoothed cell count with 257 species, and Palau has the highest raw cell count with 183 species. Relatively better sampling in the Ryukyus and weaker sampling in the western Indo-Australian Archipelago is most probably reflected in the cell totals for these regions. Diversity decreases eastward through Polynesia, with regionally high smoothed cell values at Guam (151 species), Wake Island (140 species), Tahiti (148 species) and in the Tuamotu Archipelago (117 species). Nevertheless, only 87 species are recorded in Hawaii.

There are three hotspots in the Indian Ocean. The first is around Madagascar, on the islands to the east and north, and in East Africa. The maximum smoothed value of 168 species near Durban surely reflects intensive sampling in this area. The next most dense cells are 114 species on Zanzibar and 113 species in the Comores Archipelago. The second hotspot is the Red Sea with smoothed cells containing up to

102 species at the north end. The third hotspot spans the Laccadive–Maldive–Chagos–Sri Lankan regions and contains smoothed cell values up to 108 species in the northern Maldives. Perhaps due to weak sampling, the eastern shore of the Indian Ocean formed by Sumatra has low species cell values. Finally the Caribbean has moderate hotspots of 94 species in Florida and 84 in Jamaica.

(b) Latitudinal clines

Latitudinal gradients in diversity, noted in many other taxonomic groups, are relatively well defined for reef fishes in the western Pacific. The south end of the Great Barrier Reef has 800 species of fishes (69 in our smoothed data), while the north end, closer to the equator, has 2000 species (118 in our smoothed data; Sale, 1976). Emery (1978) points out that the latitudinal gradient in diversity is correlated with a host of parameters. These include increasing seasonality and frequency of winter storms, decreasing solar radiation, the geographic area, the continuity of productivity, and the long-term climatic stability in high latitudes. In the Red Sea, on the other hand, peak fish diversity and the diversity of many other reef groups (e.g. corals and molluscs) is found in the central and northern region (Sheppard *et al.*, 1992). In the south, extreme environmental conditions and the amount of hard substrate restrict the growth of reef-building corals.

Correlations between the number of species in all families and the latitude were not significant for several partitions of the globe tested including: (1) north of the equator; (2) south of the equator; and (3) grouped poleward from the equator. Increases toward the equator are offset by a marked decrease in the bands immediately north and south of the equator. When the correlation is calculated without the two equatorial bands the correlation (-0.81) is significant ($p=0.02$). The reduced number of species near the equator is likely due to: (1) a reduction in coral reef habitat on the west coast of South America due to cool upwelling; (2) the large influx of fresh turbid water from the Amazon and Orinoco rivers on the east coast of South America; and (3) upwelling and influx of fresh turbid water from the Zaire river on the west coast of Africa. Parts of the coasts of Somalia in the Gulf of Aden and the Indian Ocean are exposed to seasonal monsoon-driven upwellings of cool water. There are coral reefs, untouched by upwelling on the Somali coast, but these are poorly sampled for coral fishes. The equatorial zone species counts are high only because of the species-rich 'coral triangle' around the East Indies.

(c) Longitudinal clines

Longitudinal clines in diversity are less well studied than latitudinal clines. However, Briggs (1986) analyzed sizable data sets on a number of biota that were grouped by biogeographic regions. We build on his approach and used the grid system to detect more precisely where

changes in diversity occur independent from biogeographic concepts. Figure 10.2 shows the changes in numbers of species around the tropics in the all-family sample. The results confirm that there are species-rich and species-poor zones and that the richest area is in the coral triangle from Sumatra east to the Marshall and Gilbert islands, with another moderately rich area in the West Indies. The GIS data enable us to quantitatively define the borders of these two areas and to define two more species-rich areas. The species-rich zones are:

(1) the coral triangle from 100 to 170°E, with species counts per 10° band ranging from 117 to 341 species and a mean count of 215 species per band;
(2) the Indian Ocean from East Africa to Sri Lanka (i.e. 30–80°E), with species counts ranging from 23 to 247, and a mean of 155 species per band;
(3) eastern Polynesia from Samoa and Hawaii east to the Society Islands (i.e. 130–170°W), with counts ranging from 56 to 165 and averaging 108 species per band; and
(4) the Caribbean Region including the West Indies, the Caribbean Sea and Gulf of Mexico (i.e. 60–100°W) with counts ranging from 88 to 129 with a mean count of 103 species per band.

The four species-rich zones have mean species counts of 103–205 species per band and are separated by species-poor bands of 3–59 species with an average of 24 species per 10° band. The species density maps in Plates 7–9 show the north–south limits as well as the east–west limits of these hotspots. We also note that bringing the north–south component into the analysis would more precisely align the eastern Polynesian hotspot with Victor G. Springer's (1982) Pacific Plate fauna.

10.3.2 Range sizes and taxonomic differences

(a) Range spans

The view has been expressed that most coral reef fish species are not threatened by human activities because populations are dispersed across wide ranges and this buffers the species against extinction from local impacts. We tested this hypothesis by calculating the frequency of range spans. Data in Table 10.3 show that 59% of the sample have range spans under 6660 km, 33% of the sample have range spans under 2220 km and 22% of the sample have range spans under 1110 km. That means that one-third of the species have range spans smaller than the length of the Great Barrier Reef (2300 km). Thus ranges of coral fishes fall into the same pattern found by McAllister *et al.* (1986) and Anderson (1984a, b), where North American freshwater fishes, amphibians, reptiles and birds are dominated by species with relatively small ranges. The hypothesis that a large range offers protection is not supported for the majority of coral reef fishes. Some of the species with small ranges will most probably be shown to have larger ranges in the future (e.g. deepwater coral fishes). These species are simply more poorly sampled than those from shallow water.

The GIS database cannot yet be used to measure the relationship between the number of species of corals and coral fishes. Harmelin-Vivien (1989) calculates this relationship for 14 Indo-Pacific sites varying from a small local reef to the Great Barrier Reef and the correlation is strong ($r=0.94$, $p <0.001$). A part of this correlation may reflect the species–area relationship due to the variation in the size of reefs. However, although this relationship holds true on the broad geographic scale, within reefs it is less obvious and the two factors are not strongly correlated (Harmelin-Vivien, 1989).

(b) Taxonomic differences in patterns

A Kolmogorov–Smirnov test of the distribution of numbers of species of chaetodontids and labrids in the 4° latitudinal bands showed no significant differences between families ($D = 0.0401$, $p = 0.4708$). A similar comparison of chaetodontids and serranids in 10° longitudinal bands was significant ($D = 0.1689$, $p < 0.0001$). One can conclude that on a global scale optimal solutions for protection of biodiversity may differ among families. The tiger reserve hypothesis is thus not supported for the oceans, confirming results for freshwater and terrestrial vertebrates by Anderson (1984a, b) and McAllister *et al.* (1986).

(c) Taxonomic diversity

The finding that higher taxon richness predicts species richness in ferns, birds and bats (Williams and Gaston, 1993) is confirmed for coral fishes. The number of smoothed coral fish species is very strongly correlated with the number of genera ($r=0.96$), and less strongly related to the number of families ($r=0.71$). The weaker relationship with families may be an artifact of the small number of coral fish families. This may also reflect differences in the geographic coverages considered in studies that treat family-level taxa.

The correlations between species, genera and families demonstrate that the number of genera or even families in an area may reflect species diversity in coral fishes. This property could be used to advantage where decisions have to made quickly on whether to save a threatened area. For example, the RAP process could be modified to make a list and count the genera of several higher taxa rather than count the species of a few taxa. This would certainly increase the efficiency of the RAP and broaden its foundation.

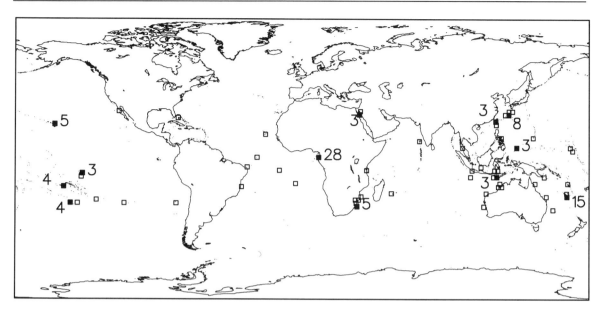

FIGURE 10.3 This is the world distribution of records of endemic species of coral fishes presented in the equidistant cylindrical projection. Open squares represent cells with one or two species and filled squares represent cells with three or more species (labeled with the species number). Endemism is defined here to include species with ranges that extend across a diameter of four cells or fewer. Approximately 92% of the coral fish endemics studied occupied a single cell. Symbols overlap since they are plotted in a slightly larger format than the 2-degree cells they represent.

10.3.3 Centers of endemism and extinction

Centers of endemism are areas where large proportions of localized endemic species are present. Our data show that 147 species or 18.4% of the 799 species in the all-family sample are endemics. Also 133 species or 16.7% occupy a single cell and 15% of the 114 chaetodontids occupy a single cell. These results indicate that one-sixth of coral fishes occupy very small ranges (i.e. on the order of 225 km wide).

Figure 10.3 shows areas of endemism world wide. A result that we did not expect is that the richest area in our database is the Gulf of Guinea on the Atlantic coast of Africa (28 species). Several other clusters of endemics in areas where there are relatively few total species include the Gulf of Guinea (28 endemics out of 42 species, 67%), New Caledonia (15

endemics out of 133 species, 11%) and Hawaii (5 endemics out of 87 species, 6%).

Many studies of endemism contain pseudo-endemics. **Pseudo-endemic** species have *known* ranges that are geographically restricted but their actual ranges are considerably larger. Pseudo-endemics are more frequent in groups that are poorly studied taxonomically and have relatively high proportions of newly described species. Pseudo-endemic species also often are from habitats that are poorly sampled. For example, coral fishes with habitats in deep water and on rough bottoms are more poorly sampled. Therefore, it is expected that these ecotypes will contain more pseudo-endemics. Fifteen percent of the relatively well-studied butterflyfishes are endemic species while 18.4% of the species in the all-family sample are endemics. This suggests that the all-families species sample is reasonably good (has few pseudo-endemics) since it does not contain a significantly higher proportion of endemics than the well-studied butterflyfishes.

Centers of endemism are largely located around isolated islands or island groups and in marginal areas (e.g. the Red Sea). Briggs (1974) lists many such marine areas of endemism. For example, Lord Howe Island, about 300 miles east of Australia, has 390 species of inshore fishes (Allen *et al.*, 1976). Of these species, 60% are wide-ranging tropical forms; 10% are endemic to Lord Howe Island, southern Australia and New Zealand; and less than 4% are endemic to Lord Howe and nearby Norfolk Island. Off Isla de Paques (Easter Island) in the eastern Pacific endemism reaches 26% (Randall and Cea Egaña, 1989). The Hawaiian Island shore fishes show 34% endemism, the eastern Pacific Cocos Island fishes show 7% endemism and 9–10% of the Revillagigedos Archipelago fish are endemic (Briggs, 1974). Approximately 10–15% of the 1000 fish species in the Red Sea are endemic (Botros, 1971).

However, these various percentages include a number of non-coral reef fish species and, thus, are not comparable with our data. In addition, our definitions of endemism are not the same. The *in situ* speciation associated with these islands is dependent upon the area, the degree of geographic isolation, the direction of the currents and the changes in climate over geological time. Centres of endemism for fishes also overlap with those for other taxa. Approximately 17% of reef fishes (by the criteria of Ormond and Edwards, 1986), 5.3% of the echinoderms and 6.3% of the corals are endemic in the Red Sea.

Centres of endemism are important for conservation because of the high proportion of unique species. The small ranges that endemics occupy make it more likely that human impacts could lead to extinction. That likelihood will vary according to the local human population size, its growth and the extent of its environment-disturbing activities. Those endemics occupying a single cell are at high risk from human impacts. A crown-of-thorns plague, a spill of toxic chemicals, or sedimentation from deforestation or agriculture may push an endemic species toward extinction. Risks of extinction are higher when an individual threat is serious, when several threats are coincident and when the threats persist, as is the case with oil spills in the Arabian/Persian Gulf (Roberts *et al.*, 1993).

Global warming poses a number of threats to the survival of species. Insular endemics are especially at risk if no island or series of islands can act as stepping stones to suitable habitat when the current range becomes uninhabitable. If there are stepping stones in the right thermal direction, survival is still not guaranteed. All the requirements of the life cycle must be met; this includes adequate food, habitat, spawning grounds and currents to carry eggs and larvae to the new island. Migration of adult coral fishes to a new island is highly unlikely for most species. The hypothesis that climatic change is likely to cause extinctions is supported by Briggs' (1966) *a posteriori* analyses of endemism in isolated oceanic islands. He ascribes the low degree of endemism found on isolated north and middle Atlantic islands to a drop in sea surface temperature that occurred during Pleistocene glaciations. Thus risks of extinction in isolated oceanic island endemics are high, whether the climate warms or cools. Springer and Williams (1990) also argue that extinctions follow cold coastal upwelling during glacially lowered sea levels in the Indo-Australian region.

GIS can be used to model the likelihood of survival for insular endemic species. For example, suppose with Peters (1988) that a 3° warming shifts the isotherms 250 km toward the poles (cells in the current grid are about 220 km wide). It should be possible to model the chances of transport of spawn to a refuge given data regarding surface currents, temperatures and reproduction activity (e.g. time of spawning, duration of egg and larval stages). Similar models calculated the likelihood of storm-blown boats carrying human cultures in different directions in Polynesia.

10.4 Conclusions

Equal-area grids and GIS are shown to be useful manual or computerized tools for studying geographic patterns in biological diversity. Numbers of species, taxa, and endemic species per area or cell are relevant to conservation planners. Similarly, patterns of human activities that impact on the environment can be localized with data stored in equal-area grids and GIS data sets. Knowledge of those patterns is useful in prioritizing areas for protection, foreseeing impacts of pending threats, and planning other preventative or remedial conservation activities.

One of the advantages of grid GIS analysis is that concepts can be tested on quantified data. This is important for research on biodiversity, biogeography and ecology. It is also an important tool for establishing thc wise use of the often limited resources available to conservationists. Hopefully the Global Environment Facility and other Earth Summit activity funders will provide significant new support for biosystematics, biological inventories and ecological studies to underpin GIS studies and conservation efforts.

10.5 Summary

Coral reefs in many areas of the world are degraded by human activities, but geographic information about the fishes (approximately 4000 species) on those reefs is unavailable or widely scattered. Based on an ongoing global study, this chapter describes the use of a GIS to assemble, analyze and present information to support the conservation of coral reef fishes. This includes the geographic occurrence, threat status and areas of high diversity and endemism ('hotspots') for these species.

An equal-area grid is used to summarize the distribution patterns of species. The grid cells are 2° latitude wide and approximately 50 000 km^2 in area. Data about the geographic distribution of 799 coral reef species in the families Gobiidae (12 species), Serranidae (264), Labridae (260), Chaetodontidae (114), Zanclidae (1), Acanthuridae (63) and Tetraodontidae (85) are compiled and analyzed. Counts of species, species within each family, genera, and families are calculated for each cell and these same sums are also calculated from the records in contiguous cell groupings. The distributions of these sum totals are then mapped.

The number of coral reef fish species increases toward the equator, but there is a drop in the number of species within 4° of the equator. It is probable that this anomaly is due to either cool oceanic upwelling or high-volume turbid river discharges. These occur near the equator, on the east and west coasts of South America, and on the west coast of Africa. Latitudinal diversity peaks near the equator in the Indo-Australian Archipelago. If the anomalous belt is excluded, there is a significant negative correlation between latitude and the number of species.

A longitudinal summary of the distribution of coral reef fishes around the globe shows four peaks of diversity: the coral triangle around the East Indies; East Africa to Sri Lanka (with three clusters); Polynesia; and the Caribbean Sea–Gulf of Mexico area. These areas average 103–205 species per 10° longitudinal band compared to an average of 24 species per 10° longitudinal band elsewhere. This spatial analysis confirms the existence of these biodiversity hotspots and helps to define their boundaries.

The longitudinal distribution of species richness for the butterflyfishes (Chaetodontidae) and seabasses (Serranidae) are analyzed using the Kolmogorov–Smirnov test. The numbers of species in these two groups are very distinctive

(D=0.1689, p<0.0001) and this demonstrates that major taxa may differ in the geographic location of their hotspots.

Family and generic richness is strongly correlated with species richness and this suggests that the locales of higher taxa can be used to rapidly assess the importance of conserving a threatened seascape. The inclusion of counts for genera and families as well as for species in an area increases the comprehensiveness of the measured biodiversity.

Our data refute the view that most coral reef fishes are unlikely to be threatened by human activities because they are widespread geographically. In our sample 59% of the species ranges span less than 6660 km and 33% of the species ranges span less than 2220 km. Endemic species (i.e. most distant records less than 4 cells *apart* ca. 1000 km) comprise 18% of all the species studied. Almost 17% of the total and 15% of the butterflyfishes are only known from a single grid cell. This suggests that a number of coral fish species with small ranges may be threatened by human activities at regional and local scales. Pseudo-endemics (species whose real range is larger than their known range) may not confound this conclusion to any extent. This contention is tested in the butterflyfishes where pseudo-endemics are expected to be rare. The ratio of small-range to large-range butterflyfishes is not significantly different from that in the all-family sample.

A significant positive correlation between the number of coral species and the number of coral reef fish species in the Indo-Pacific sample lends credence to the view that the deteriorating condition of coral reef habitat poses a threat to the reef fishes. Data on anthropogenic threats will be added to the GIS data set to evaluate these impacts.

The equal-area grid technique is a useful tool for the compilation, analysis and presentation of data important to conservation assessment. It can be used to define and locate areas with high overall species richness, many endemic species, higher taxon richness, intense human environmental impacts, or with a need for increased biotic sampling. The accuracy of the results will be moderated by the level of taxonomic knowledge and the thoroughness of biological inventories. Thus taxonomic research, biological surveys, natural history museum collections and the resulting authoritative spot distribution maps are important resources for carefully planned conservation.

Acknowledgements

We gratefully acknowledge funding support for the GIS project from The Curtis and Edith Munson Foundation of Chicago, the Sir Peter Scott Trust for Education and Research in Conservation, the Sir Peter Scott IUCN/SSC Action Plan Fund, the British Ecological Society/Coalbourn Trust, The Norcross Wildlife Foundation and Ocean Voice International. Susan Tressler and Dr George Rabb of the Species Survival Commission provided invaluable assistance in finding support. The Canadian Museum of Nature provided computer time, office space, and library and other resources. The Department of Marine Sciences and Coastal Management of the University of Newcastle-on-Tyne, and the Eastern Caribbean Center of the University of the Virgin Islands, provided office and library facilities for compiling maps.

The late Sir Peter Scott, Gustavo Nunan, and the many other people who have generously donated their records of fish species distributions are gratefully acknowledged for the information that provided the kernel of our database. The second author's father, the late F. William Schueler III, helped develop the formula for the equal-area cells. Noel Alfonso and Ian Jones spent weeks drafting the original equal-area basemaps and compiling data on occurrence of genera of reef-building corals. Noel Alfonso also reviewed the manuscript.

Brian Eddy provided advice and training on use of the GIS package.

References

Alcala, A.C. and Gomez, E.D. (1987) Dynamiting coral reefs for fish: A resource-destructive fishing method, in *Human Impacts on Coral Reefs: Facts and Recommendations* (ed. B. Salvat), Antenne Museum, E.P.H.E., Tahiti, pp. 51–60.

Alfonso, N. (1991) An equal-area grid map for GIS: A tool for biodiversity conservation and biogeography. *Canadian Biodiversity*, 1(1), 30–2.

Allen, G.R. (1975) *Damselfishes of the South Seas*, T.F.H. Publications, Inc., Neptune City, New Jersey, 240pp.

Allen, G.R., Hoese, D.F., Paxton, J.R. *et al.*, (1976) Annotated checklist of the fishes of Lord Howe Island. *Records of the Australian Museum*, 30, 365–454.

Anderson, S. (1984a) *Geographic Ranges of North American Birds*, American Museum Novitates (2785), pp. 1–17.

Anderson S. (1984b) *Areography of North American Fishes, amphibians and reptiles*, American Museum Novitates (2802), pp. 1–16.

Ansula, A.C. and McAllister, D.E. (1992) Fishing with explosives in the Philippines. *Sea Wind*, 6(2), 6–12.

Botros, G.A. (1971) Fishes of the Red Sea. *Oceanography and Marine Biology Annual Review*, 9, 221–348.

Briggs, J.C. (1966) Oceanic islands, endemism, and marine paleotemperatures. *Systematic Zoology*, 15(2), 153–63.

Briggs, J.C. (1974) *Marine Zoogeography*, McGraw-Hill, New York, 475pp.

Briggs, J.C. (1986) Species richness among the tropical shelf regions, in English translation of *Soviet Journal of Marine Biology*, Plenum Publishing Corporation, pp. 295–302.

Clarke, A. (1992) Is there a latitudinal diversity cline in the sea? *Trends in Ecology and Evolution*, 7(9), 286–7.

D'Elia, C.F. Buddemeier, R.W. and Smith, S.V. (1991) *Workshop on Coral Bleaching, Coral Reef Ecosystems and Global Change: Report of the Proceedings*, Maryland Sea Grant Publication UM-S4-TS-91-03. 49pp.

Dubinsky, Z. (ed.) (1990) *Coral reefs. Ecosystems of the World 25*, Elsevier, Amsterdam, Oxford, 550pp.

Earth and Ocean Research Limited (1991) *inFOcus: A Geographic Information Management and Display System*, Earth and Ocean Research Ltd, Dartmouth, Nova Scotia: users guide, complex pagination.

Emery, A.R. (1978) The basis of fish community structure: Marine and freshwater comparisons. *Environmental Biology of Fishes*, 3, 33–47.

Environmental Sciences Limited (1990) *QUIKMap Version 2.50 Users Guide*. Environmental Sciences Limited, Sidney, British Columbia: complex pagination.

Fox Software, Inc. (1991) *FoxPro*, 3 vols, Fox Software, Perrysville, Ohio.

Glynn, P.W. (1991) Coral reef bleaching in the 1980s and possible connections with global warming. *Trends in Ecology and Evolution*, (6), 175–9.

Gomez, E.D. and Alcala, A.C. (1979). Status of Philippine coral reefs 1978. *Proceedings of the International Symposium on Marine Biogeography and Evolution of the Southern Hemisphere*, 2, 663–9.

Goreau, T.J. (1990) Coral bleaching in Jamaica. *Nature*, 343, 417.

Groombridge, B. (1992). *Global Biodiversity, Status of the Earth's Living Resources*. A report compiled by the World Conservation Monitoring Centre. Chapman & Hall, London, 585pp.

Harmelin-Vivien, M.L. (1989) Reef fish community structure: An Indo-Pacific comparison, in *Vertebrates in Complex Tropical Systems* (ed. M.L. Harmelin-Vivien and F. Bourlière), Springer-Verlag, New York, pp. 21–60.

Harrison, P. (1992). *The Third Revolution, Environment, Population and a Sustainable World*, Tauris, London, New York, 359pp.

ICBP (1992) *Putting Biodiversity on the Map: Priority Areas for Global Conservation*, International Council for Bird Preservation, Cambridge, UK.

Magurran, A.E. (1991) *Ecological Diversity and its Measurement*, Chapman & Hall, London, 179pp.

McAllister, D.E. (1988) Environmental, economic and social costs of coral reef destruction in the Philippines. *Galaxea*, 71, 161–78.

McAllister, D.E. (1991a) What is the status of the world's coral reef fishes? *Sea Wind*, 5, 14–18.

McAllister, D.E. (1991b) What is biodiversity? *Canadian Biodiversity*, Ottawa, 1(1), 4–6.

McAllister, D.E. (1993). The eco-taxa hotspot approach to the conservation of biodiversity: Evolving tools. *Association of Systematics Collections Newsletter*, 21(4), 47–8.

McAllister, D.E., Platania, S.P., Schueler, F.W. *et al.* (1986) Ichthyofaunal patterns on a geographic grid, in *Zoogeography of North American Freshwater Fishes* (ed. C.H. Hocutt and E.O. Wiley), Wiley, New York, pp. 18–51.

McManus, J.W. (1988) Coral reefs of the ASEAN region: Status and management. *Ambio*, 17, 189–93.

McNeely, J.A., Miller, K.R., Reid, W.V. *et al.* (1990) *Conserving the World's Biological Diversity*, IUCN,

WRI, CI, WWF-US and World Bank, Gland, Switzerland, 193pp.

Myers, R.F. (1989) *Micronesian Reef Fishes*, Coral Graphics, Guam, 298pp.

Newell, N.D. (1971) *An Outline History of Tropical Organic Reefs*, American Museum Novitates (2465), pp. 1–37.

Norris, E.A. (ed.) (1993) *Global Marine Biological Diversity. A Strategy for Building Conservation into Decision Making*. Island Press, Washington, D.C., 383pp.

Ormond, R.F.G. and Edwards, A.J. (1986) Red Sea fishes, in *Red Sea* (ed. A.J. Edwards and S.M. Head), Pergamon Press, Oxford, UK, pp. 251–87.

Peters II, R.L. (1988) The effect of global climatic change on natural communities, in *Biodiversity* (ed. E.O. Wilson), National Academy Press, Washington, D.C., pp. 450–61.

Planck, J., McAllister, D.E. and McAllister, A. (1988) *Shiraho Coral Reef, and the Proposed New Ishigaki Island Airport, Japan, with a Review of the Status of Coral Reefs of the Ryukyu Archipelago, Japan*. Species Survival Commission, International Union for the Conservation of Natural Resources, Switzerland, 232pp.

Randall, J.E. (1980) Conserving marine fishes. *Oryx*, 15(3), 287–91.

Randall, J.E. (1982) Tropical marine sanctuaries and their significance in reef fisheries research. Proceedings of a workshop held October 7–10, 1980 at St Thomas, Virgin Islands of the United States (ed. G.R. Huntsman, W.R. Nicholson and W.W. Fox, Jr). The Biological Bases for Reef Fishery Management. NOAA Technical Memorandum NMFS-SEFC-80, pp. 167–78.

Randall, J.E. and Cea Egaña, A. (1989) *Canthigaster cyanetron*, a new toby (Teleostei; Tetraodontidae) from Easter Island. *Revue francaise d'Aquariologie*, 15(3), 93–6.

Roberts, C.M. (1987) *Report on the Distribution of Coral Reef Fish Species throughout the Indo-West Pacific Region using Data Collected by Sir Peter Scott*, Tropical Marine Research Unit, York, UK.

Roberts, C.M. and Hawkins, J.P. (1992), Mapping the distribution of coral reef fishes. *Sea Wind*, 5(4), 3–8.

Roberts, C.M., Downing, M., and Price, A.R.G. (1993) on troubled waters: Effects of the Gulf War on coral reefs, in Proceedings of a meeting entitled 'Global Aspects of Coral Reefs: Health Hazards and History' held at the Rosentiel School of Marine and Atmospheric Sciences, Miami, Florida, USA, June 1993, pp. V35–V41.

Robinson, S. (1987) The killing of Costa Rica's Caribbean coral reefs. *Sea Wind*, 1(2), 8–15.

Rubec, P.J. (1986) The effects of sodium cyanide on coral reefs and marine fishes in the Philippines, in *The First Asian Fisheries Forum* (ed. J.L. Maclean, L.B. Dizon and L.V. Hosillos), Asian Fisheries Society, Manila, Philippines, pp. 297–302

Sale, P. (1976) Reef fish lottery. *Natural History*, 85, 60–5.

Sale, P. (ed.) (1991) *The Ecology of Fishes on Coral Reefs*, Academic Press, San Diego, 784pp.

Samoilys, M.A. (1988) Abundance and species richness of coral reef fish on the Kenyan coast: The effects of protective management and fishing. *Proceedings of the 6th International Coral Reef Symposium, Townsville*, 2, 261–6.

Schueler, F.W. (1982) *Geographic Variation in Skin Pigmentation and Dermal Glands in the Northern Leopard Frog*, Rana pipiens. Publications in Zoology, National Museum of Natural Sciences, National Museums of Canada 16, pp. 1–80.

Sebert, L.M. (1975) *The Mercator and Transverse Mercator Projections. Energy, Mines and Resources Canada*, Technical Report No. 69–1, 12pp. Surveys and Mapping Branch.

Sheppard, C.R.C., Price, A.R.G. and Roberts, C.M. (1992) *Marine Ecology of the Arabian Region: Patterns and Processes in Extreme Tropical Environments*, Academic Press, London, 359pp.

Smith, S.V. (1978) Coral reef area and the contributions of reefs to processes and resources of the world's oceans. *Nature*, 273, 225–6.

Smith, S.V. and Buddemeier, R.W. (1992) Global change and coral reef ecosystems. *Annual Reviews in Ecology and Systematics*, 23, 89–118.

Sokal, R.R. and Rohlf, F.J. (1981) *Biometry*, Freeman, New York, 859pp.

Springer, V.G. (1982) *Pacific Plate Biogeography, with Special Reference to Shore Fishes*, Smithsonian Contributions to Zoology (367), pp. 1–182.

Springer, V.G. and Williams, J.T. (1990) Widely distributed Pacific Plate endemics and lowered sea-level. *Bulletin of Marine Science*, 47(3), 631–40.

Thresher, R.E. (1992) Geographic variability in the ecology of coral reef fishes: Evidence, evolution, and possible implication, in *The Ecology of Fishes on Coral Reefs* (ed. P. Sale), Academic Press, San Diego, pp. 401–36.

UNEP/IUCN (1988) *Coral Reefs of the World*. Hardcovers: Vol. 1, *Atlantic and Eastern Pacific*, 373 pp.; Vol. 2, *Indian Ocean, Red Sea and Gulf*, 389pp.; Vol. 3, *Central and Western Pacific*, 329pp. UNEP Regional Seas Directories and Bibliographies; IUCN, Gland, Switzerland and Cambridge, UK UNEP, Nairobi, Kenya.

Vane-Wright, R.I., Humphries, C.J. and Williams, P.H. (1991) What to protect? Systematics and the agony of choice. *Biological Conservation*, 55, 235–54.

Vernon, J.E.N. (1986) *Corals of Australia and the Indo-Pacific*, Angus and Robertson Publishers, London, 644pp.

Vernon, J.E.N. (1993) *A Biogeographic Database of Hermatypic Corals*, Australian Institute of Marine Science Monograph Series 10, pp. 1–433.

Weber, P. (1993) Reviving coral reefs, in *State of the World 1993* (ed. Lester R. Brown), W.W. Norton, Washington, pp. 42–60.

Weller, W.F. and Oldham, M.J. (1988) *Ontario Herpetological Summary 1986*, Ontario Field Herpetologists, Cambridge, Ontario, 221pp.

Wells, S. and Hanna, N. (1992) *The Greenpeace Book of Coral Reefs*, Blandford, London, 160pp.

Williams, P.H. (1993) Choosing conservation areas: Using taxonomy to measure more of biodiversity, in *Biodiversity Conservation* (ed. T.-Y. Moon), Korean Entomological Institute, Seoul, pp. 194–227.

Williams, P.H. and Gaston, K.J. (in press) Measuring more of biodiversity. Can higher taxon-richness predict wholesale species richness? *Biological Conservation*.

Williams, P.H. and Humphries, C. (1992) Biodiversity taxonomic relatedness and endemism in conservation, in *Systematics and Conservation Evaluation* (ed. P.L. Forey, C.J. Humphries and R.I. Vane-Wright), The Systematics Association, Oxford University Press, pp. 1–14.

Woodland, D.J. (1986) Wallace's Line and the distribution of marine inshore fishes, in *Indo-Pacific Fish Biology. Proceedings of the 2nd International Conference on Indo-Pacific Fishes* (ed. T. Uyeno *et al.*), Ichthyological Society of Japan, Tokyo, pp. 453–60.

Part Seven

A Continental Conservation Monitoring Program

The Environmental Resources Information Network (ERIN) in Australia is well advanced in developing a continental environmental spatial information system to underpin the development and implementation of environmental policy. This system is designed around key patterns and processes in Australian landscapes. A primary project, presented in Chapter 11, is the coordinated development of data sets for plant species that dominate Australian vegetation. Collaborative efforts are integrating collections of plant information from across Australia, leading to the development of a national plant list and giving Australia the capacity to effectively analyze and monitor its biodiversity. Spatial information systems areused to link these and other data sets with bioclimatic modeling and remote sensing to provide an integrated framework for environmental decision making.

The Australian system presented in Chapter 11 represents a well-maintained, continental approach to monitoring plants and animals with databases and maps. The evolving Australian environmental information system includes the important capacity to produce regional patterns at varying scales. In the future, this system anticipates establishing a network of distributed databases across the Australian continent. Chapter 11 presents a reassuring picture of an existing monitoring system that is perfecting the use of maps and databases for conservation into the future.

Chapter eleven

Linking plant species information to continental biodiversity inventory, climate modeling and environmental monitoring

Arthur D. Chapman and John R. Busby

11.1 A continental environmental information infrastructure

11.1.1 Introduction

The future productive capacity of the world's lands and seas and the quality of life of its people ultimately depend upon the integrity of the natural environment. The development of environmental policies and sound resource management practices depend on the ready availability of high-quality data on a wide range of environmental issues. The decision-making process is severely constrained by de-

Mapping the Diversity of Nature. Edited by Ronald I. Miller.
Published in 1994 by Chapman & Hall, London. ISBN 0 412 45510 2.

ficiencies in both the quality and availability of relevant information. These constraints reduce our capacity to adequately explore options for ecologically sustainable development.

In recognition of this information deficiency, the Australian Government in 1989 established the Environmental Resources Information Network (ERIN). ERIN has the mission 'To provide geographically related environmental information of an extent, quality and availability required for planning and decision making'. To fulfill this mission, ERIN is building a continental-scale spatial information system to focus upon key attributes that reflect environmental patterns and processes. It is also developing collaborative and physical networks to enhance the flow of information between data providers and information users. These users include policy makers, land managers and environmental researchers.

ERIN is linked to the international data network, Internet, and is developing methods of making environmental information available over the wide area network. A prime objective of ERIN is for custodial institutions to maintain their own data. The goal envisages queries being made via transparent data links over a network of distributed databases with links maintained through ERIN. The goal is not yet attainable as the necessary technology is not available and many Australian institutions holding data are yet to be linked to a network.

ERIN has set a number of strategic directions (Miles, 1992), including:

- the use of primary data
- integration of modeling tools
- development of approaches for identifying and characterizing regional environmental patterns for use in environmental assessment and planning
- development of broad-based collaborative projects.

Since the expertise available within any one agency can never be capable of covering all the issues involved, ERIN is investing considerable effort into developing collaborative and physical networks with agencies across the continent and around the world. This collaboration facilitates access to the best available expertise and information and will allow ERIN, in the long-term, to address a multiplicity of environmental issues.

11.1.2 Data considerations

ERIN provides products which display data at many different spatial scales. The ERIN system is based upon point-referenced data which largely eliminates problems related to spatial scale. Data collected and referenced to points can be analyzed and the results displayed at any scale. For example, a continental scale vegetation map most appropriately displays the cover and height of prominent vegetation layers whereas a catchment map presents more comprehensive floristic community information. Data preclassified at the time of collection cannot be used to produce both types of map. Both, however, can be produced using primary data stored as point locations.

In the ERIN system, primary data refer to those data which underpin all the data types similar to the elements in a chemical compound. These data include individual point-referenced specimens of flora and fauna with associated attribute information (e.g. currently ascribed scientific name, collector's name, date of collection, locality and altitude). The objective is to minimize limitations that are imposed by map scale and the use of preclassified entities.

Preclassified entities commonly used to produce natural resource maps include: vegetation categories, soil types, tree height classes and species (i.e. a collection of individual specimens). Preclassified data, however, present problems: when concepts underpinning the classification change, the individual categories also change and thus the underlying data become valueless. Data stored as primary attributes (e.g. individual specimens, actual tree heights, etc.) can be used to produce classified entities for display and communication while remaining available for use in alternate classifications.

ERIN has developed the capacity to identify and characterize regional environmental patterns that involve the classification of the land and water into categories based on similar environmental attribute values. This capacity to generate regional patterns at varying scales (or regionalizations) has allowed ERIN to provide a valuable context for environmental assessment and planning within Australia. The resulting maps, which are based on primary attribute data, can be tailored to user requirements (Thackway, 1992), can be produced at a variety of scales and can incorporate any number of alternative classifications.

11.1.3 Systems infrastructure

The hardware and software infrastructure that supports the ERIN system requires considerable intellectual and financial commitment. A major investment has been made in developing integrated database modules to handle large volumes of plant and animal specimen data, associated taxonomic and management information, and information on the databases themselves.

Each of the modules is linked in the ERIN system and each is outlined below.

1. The *Taxon* module manages taxon names and allows for easy updating of names as taxonomic revisions are completed. This module includes records of plant names from the Census of Australian Vascular Plants and the Australian Plant Name Index and vertebrate animal names from the Census of Australian Vertebrate Species. All of these databases are maintained by agencies outside of the core Unit establishing the ERIN system.

2. The *Specimen* module manages individual records of plant and animal specimens. The data elements include for each record: the custodial institution name and institutional identifier, a taxon name, the location (geocoded by latitude and longitude and including a precision estimate), the altitude with a precision estimate, the collector's name, and the collection date.

3. The *Data Dictionary and Catalogue* module is used to document and manage data sets and allows for the automatic addition of data source citations to maps produced by the GIS. The module includes information on custodianship, data restrictions, and quality.

11.1.4 Objectives

ERIN considers Spatial Information Systems (SISs) as much enhanced developments of Geographic Information Systems (GISs) that otherwise represent only an automated map production capacity. In addition to a GIS, a SIS needs to incorporate appropriately designed relational databases, modeling and analytical tools and collaborative and physical networks to link environmental data and expertise for use by managers and decision makers. The modeling and analytical tools provide the capacity for managers to explore options, to identify gaps in the data and to reach sound decisions about managing the environment.

11.2 Integrating Australia's species information

11.2.1 Introduction

Databases of plant and animal specimens have existed for centuries in the form of dried plant specimens in herbaria and preserved specimens in museums. These invaluable collections are a vital information source for taxonomic studies and information on the historical ranges of species now restricted by habitat modification. These collections, however, have been of minimal value for environmental assessment and management. Manual processing of records to map the distribution of even a single taxon in a designated area was inordinately time consuming. Computer processing and the ability to inexpensively store large volumes of data makes this practical, not only for single taxa but for large numbers of taxa, and thus dramatically enhances the value of these collections (Chapman, 1992).

11.2.2 Data capture and standardization

It is important to maximize the use of existing data since the resources to recollect data with traditional surveys and research are severely limited (Bridgewater, 1991). It costs about $6 Australian to database a herbarium or museum record. It costs several times that amount to collect the specimen in the field, particularly if this involves paid professional staff. New surveys require careful planning to maximize the use of the existing information base and of the limited available resources.

It is also vital to establish and adhere to standards for recording survey information. Standards are the means by which information is communicated between people and are thus vital in bringing together all the relevant interest groups to address an environmental issue. Standards include the selection of attributes that represent a phenomenon, the meaning of these attributes (i.e. their characteristics and range of values), and the methods for communication of these attributes. ERIN is fostering the development of standards for a diverse range of environmental features in collaboration with all the major interest groups. These include a set of core attribute standards for biological site records (Bolton, 1992) and standards for the interchange of herbarium specimen information (Croft, 1988).

11.2.3 Establishing priorities

The major Australian collecting institutions currently hold some 5 million plant specimens and 29 million animal specimens or 'lots'. 'Lots' are produced during collections of large numbers of specimens for a species during one brief period at one site (Richardson and McKenzie, 1991). The plant specimens are scattered across some 13 institutions and the animal specimens across nine institutions. Only 19% of the plant specimens and 8% of the animal specimens are currently stored in databases. These collections are currently increasing at a rate of ca. 5% per year and, with the present rate of data capture, it will take many decades for the backlog to be put into databases. In addition, several million observational records are held in universities, government nature conservation agencies and land management agencies.

It is possible, of course, to establish and implement priorities for the computer capture of much of this data backlog. ERIN has set priorities for the capture of this information as part of its mission to develop a data infrastructure for environmental management and decision making. First priority is given to the acquisition of data on endangered and vulnerable plant and animal species. Also included are those plant species prominent in the canopies of Australian vegetation types, the 'land cover' species. Large existing data sets on birds, bats, koalas and kangaroos, which can be acquired at little cost, are also being brought into the network.

11.2.4 Australian Landcover Species Project

The ERIN Landcover project aims to collate data on plant species which comprise the various landcover elements in the Australian landscape (Chapman, 1991a). Broad-based collaborative arrangements are developed whereby ERIN provides assistance for the capture and upgrade of distributional data and in return receives a copy of these data. Arrangements are made with custodial agencies whereby they retain ownership over the data. ERIN retains the prerogative to aggregate, enhance or analyze the data for its own purposes, but undertakes not to pass the original data to a third party without permission from the custodian.

The project aims to collate the data under three time frames:

- **Short term.** Data that can be assembled over a 12-month period (this usually includes data already in a database).
- **Intermediate.** Data that can be put into a database over about an 18-month period with only a small investment of funds.
- **Long term.** Development of collaborative arrangements to put into databases whole taxonomic groups of plants. This also includes the development of standards for the capture and interchange of the data.

A major aim of the project is to channel the country's effort into putting data of entire collections of plant taxonomic groups into databases on a national basis. Until now, each institution focused upon different plant taxonomic groups, with the result that no one group of plants was available for the whole continent. This has contributed to the difficulties associated with the use of these data for environmental management or continent-wide biodiversity assessment.

Important landcover elements in Australia include the genus *Eucalyptus*, arid species of grasses, members of the genera *Acacia* and *Callitris*, the Chenopodiaceae and Casuarinaceae. ERIN databases now include data on over 200 000 *Eucalyptus* records and 120 000 grass records.

The Taxon module in the ERIN database allows users to distinguish and assemble names of diverse sets of taxa and thus to extract data easily for these groups. This is particularly useful where the group of plants or animals of

interest includes representatives in several taxa. Categories of taxa that have been set up in ERIN include mangroves, C3 and C4 grasses and mallee species of eucalypt. The allocation of species to these categories is implemented by an 'Applied Categories' attribute table in the database. This allocation capability has been used in the preparation of a comprehensive mangrove coverage for Australia for use in coastal zone management and planning.

11.2.5 Endangered and vulnerable plants and animals

Australia publishes lists of nationally extinct, endangered and vulnerable plants and animals (Leigh and Briggs, 1992; ANZECC, 1991). These lists help to determine national priorities for research, management and conservation and for the development of recovery plans. Data on endangered and vulnerable plant and animal species in Australia have been collated over a 10-year period and are loaded in the ERIN database. A modeling tool BIOCLIM (Busby, 1991; and below) is being used to predict where suitable climates for these species are located and hence where further populations may occur. This information may assist in the allocation of resources for future survey efforts, for reserve and wildlife corridor assessments and for conservation planning in regard to global climate change.

11.3 Climate analysis and distribution modeling

The present rate of gathering precise and accurate distribution data on the Australian flora and fauna will not provide a comprehensive understanding of biodiversity patterns for several decades. In the interim, surrogates are needed to detect biodiversity 'hotspots', plan reserve networks and focus new field survey efforts. Predictions of likely occurrences of a species in unsurveyed areas can be derived from correlations between known species' records and environmental factors, particularly climate. In Australia, good estimates of long-term average climate are available for the continent. These data are somewhat less reliable in remote areas and in areas with complex climate gradients where the measurement networks are inadequate. Other environmental data sets are also being compiled. BIOCLIM is a distribution modeling system in widespread use in Australia. It has recently been extended to New Zealand and Papua New Guinea. BIOCLIM is a bioclimate analysis and prediction system that is used to stratify areas on a climatic basis and to predict distributions of individual entities such as species or vegetation types. BIOCLIM is based on continuous surfaces (Laplacian smoothing spline functions) that are fitted to measured meteorological data. From inputs of latitude, longitude and altitude, BIOCLIM can be used to generate estimates of monthly mean minimum and maximum temperatures and precipitation for any point on or near mainland Australia and Tasmania. Analyses can then be based upon any attribute derivable from monthly mean climate values (e.g. mean annual temperature, mean winter precipitation). Predictions of entity distributions can be made at a variety of scales and resolutions down to about 2–5 km or the ca. 1:100 000 map scale. Finer resolutions may be possible (M.F. Hutchinson, pers. comm.). Climate estimates from points on a survey grid are matched against an entity's climate profile. This profile is produced from climate estimates of individual sites at which the entity is recorded. Differing degrees of correspondence between the grid points and the profile are used to provide individual predictions for various parts of the survey area. In addition, predictions from a number of entities are combined to highlight regions potentially containing high biological diversity (Busby, 1991).

BIOCLIM is successfully used with data for numerous Australian animal and plant species.

Published examples include C3 and C4 grasses (Prendergast and Hattersley, 1985), temperate rainforest tree species (Busby, 1986a; Gibson, 1986; Hill *et al.*, 1988), weeds (Panetta and Dodd, 1987) and small mammals (Lindenmeyer *et al.*, 1990, 1991). The most comprehensive application of BIOCLIM to date is for the 77 Australian species of snakes in the family Elapidae (Longmore, 1986).

Other published BIOCLIM applications include intercontinental climatic analyses of eucalypt plantation success (Booth, 1985; Booth *et al.*, 1987), estimations of Holocene climates (Markgraf *et al.*, 1986; McKenzie and Busby, 1992) and prediction of impacts of climate change on elements of the Australian flora and fauna (Busby, 1988; Bennett *et al.*, 1991). The capacity of BIOCLIM to support preliminary assessments of likely sensitivities of the flora and fauna to climate change will prove invaluable in initial planning for reserve networks and migratory corridors.

11.4 Multiple scale plant distribution patterns

ERIN's role is to provide infrastructure support. Accordingly, ERIN collates key strategic data sets required by a range of research groups. These data underpin projects that analyze Australian biodiversity patterns at different scales. The collaboration of ERIN staff on these projects ensures that the analysis results are made widely available to environmental management agencies. Staff also conduct internal analyses of the collated data to develop plans that address data gaps and deficiencies.

11.4.1 Using classified data

Some preliminary analyses were conducted on the information compiled in the Census of Australian Vascular Plants (Hnatiuk, 1990). The Census lists all the vascular plant species in

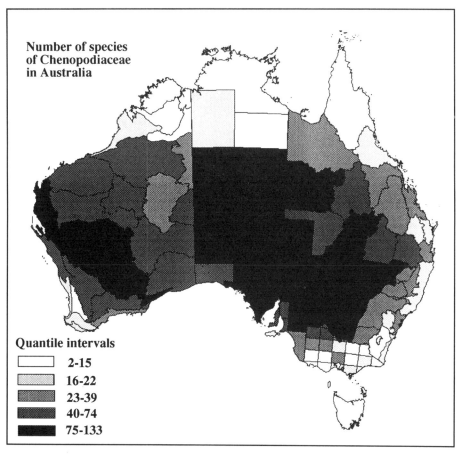

FIGURE 11.1 Pattern of distribution of Chenopodiaceae in Australia showing the number of species in each biogeographic region. Data and regions derived from Hnatiuk (1990).

Australia and their occurrences within a set of defined bioclimatic regions. These data were used to identify broad patterns of biodiversity in Australia within each of a number of the more important families and genera. Direct comparisons between the regions cannot be made because the size of the regions chosen in this analysis vary considerably. Broad patterns of diversity density, however, are recognized. For example, Figure 11.1 shows that the main concentration for Chenopodiaceae is in the arid areas of Australia. In contrast, the main concentration for Proteaceae (Figure 11.2) is in the southwestern and eastern parts of the conti-nent. This type of information may facilitate decisions about broad scale priorities, but these data are of little value for on-site biodiversity management or for reserve selection. This limitation reinforces the value of using primary data wherever possible.

11.4.2 Using point-referenced primary data

Specimen data collected under the Landcover project allows a higher resolution analysis of areas of high species diversity. For example, with the genus *Eucalyptus*, an analysis of the number of species per one degree grid square

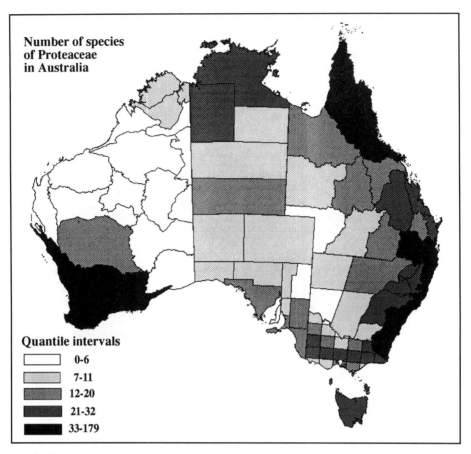

FIGURE 11.2 Pattern of distribution of Proteaceae in Australia showing the number of species in each biogeographic region. Data and regions derived from Hnatiuk (1990).

(Figure 11.3) produces a more detailed picture of areas that may be important for biodiversity management or for consideration in reserve selection. Point data, such as that used in this analysis, accommodates a virtual infinite number of data uses. GIS technology allows the data to be overlaid on different regional configurations of remotely sensed data to examine environmental interrelationships. These procedures are applied in the western portion of Victoria using a composite NOAA AVHRR image. The image is overlaid with point records of the malleefowl (*Leipoa ocellata*), a large ground-nesting bird found in the southern half of Australia, and the predicted distribution of the species are derived from BIOCLIM (Plate –9). These procedures identify areas of remaining mallee vegetation within the range of its predicted suitable climate where the malleefowl may nest. These results may be used to determine conservation sites or to identify areas in need of further survey.

11.5 Data quality assurance

11.5.1 Introduction

One of the problems of many biological data is their poor quality. Historical collections of

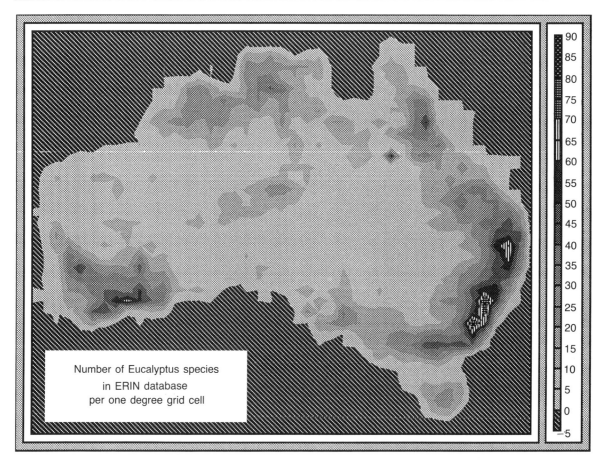

FIGURE 11.3 Plot of the number of species of the genus *Eucalyptus* in each one-degree by one-degree grid cell in Australia.

specimens, such as those from herbaria and museums, were not collected with multiple uses in mind. Thus, many attributes are inaccurately recorded or are not recorded at all. On the other hand, a different set of problems exist with survey data. The records are seldom vouchered and therefore cannot be verified.

Data need to be accurate if they are to be used for important environmental decisions. For example, the placement of a reserve to protect a particular endangered species requires accurate data (Chapman, 1992). Thus the data need to be carefully validated and checked for reliability. The ERIN landcover and endangered species projects have in excess of one million records already in the database. One cannot check each record individually, so techniques have been developed to identify 'suspect' records automatically.

11.5.2 Taxon validation

Australia is in a unique position in that it has both a comprehensive list of all names applied to Australian vascular plants (Chapman, 1991b–e) and a comprehensive Census of accepted names for vascular plants (Hnatiuk, 1990). These two databases are now combined in a single comprehensive database of accepted names and their synonyms. The current database also includes names that are accepted but as yet unpublished. These are either stored as manuscript names or names in a format agreed upon by all the major Australian botanical institutions. The creation of a similar database is now being considered for the Australian fauna. The fauna database would be created by combining records from the published volumes of the *Zoological Catalogue of Australia* (e.g. Cogger, 1983).

All species names associated with specimens or survey records are checked against the names in the Taxon module as they are loaded into the database. Records with no match are identified and returned to the custodians for checking. Names in an incoming data set that are not regarded as current are altered to reflect the current taxonomy. This is a complex problem since a name change could involve a concept change which is difficult to characterize in a database. A flag is placed in the database to indicate a taxonomic problem when it is not possible to identify a 'non-current' name. One may decide to exclude these specimens from further analysis, depending on the type of analysis to be carried out. This validation procedure ensures the validity of all the taxonomic names used by ERIN.

11.5.3 Geographic validation

(a) Suspect records

Many herbarium and museum specimens carry very little geographic information, other than a general description of the location where they were collected. Geographic codes (e.g. latitude or longitude) that define data records for use in distributional studies are seldom given. It is crucial that these records be present for any analysis involving geographic distributions.

Data that cannot be geographically coded produce various kinds of error. For example, many of the place-names used in historical collections no longer exist. In addition, many collections only include very broad locality information (e.g. 'Nova Hollandia', an early name for Australia or 'Australia Felix', an early name for a large area in southern Australia). The one place-name may refer to several different localities and it is easily misapplied. For example, there are hundreds of 'Stony Creek's in Australia. It is easy for inexperienced recorders to read one of these maps incorrectly. For example, there is often confusion in the interpretation of different coordinate systems (e.g. UTM and latitude/longitude). Another very common error is when a recorder or data-entry operator accidentally swaps numerals.

The cost to input a collection into a database can be substantial (Armstrong, 1992). Few institutions can afford to carry out further detailed checking of data once they are input into a database. Thus the accuracy of geographic information is often not checked. Tests carried out at ERIN on biological data received from both herbaria and museums indicates considerable error in supplied geographic information (up to 18% in some cases). This clearly shows some of the difficulties involved when handling geographic information from historical collections. So how should one try to identify these 'suspect' records?

(b) Validation procedures

ERIN has developed several procedures for checking geographic information. A GIS is used to check the geographic codes with broad geographic features such as the Australian coastline. This checking procedure presents few problems as it is carried out automatically by the GIS.

The most innovative and successful method, however, is the detection of outliers using species climate profiles generated by the BIOCLIM system. Statistical jack-knifing procedures

TABLE 11.1 A temperature climate profile for *Eucalyptus robusta*. The profile was produced from BIOCLIM and is based on 100 specimens of *E. robusta* from the Queensland Herbarium

	Climate variation[a]							
	1	*2*	*3*	*4*	*5*	*6*	*7*	*8*
Mean	20.3	9.3	28.7	19.4	15.6	24.2	23.8	17.0
Std deviation	1.6	2.3	1.5	1.9	1.9	1.3	1.7	2.1
Minimum value	16.8	4.9	25.2	14.4	11.5	20.9	15.2	12.4
5th percentile value	17.5	6.1	26.5	17.2	12.4	21.9	20.6	14.5
25th percentile value	19.9	7.8	28.5	18.0	14.9	24.0	23.9	15.7
75th percentile value	20.9	11.1	28.9	20.5	16.6	24.4	24.4	18.8
95th percentile value	22.5	12.4	30.6	21.4	18.0	26.1	25.9	19.1
Maximum value	26.0	18.9	35.5	26.7	24.0	27.8	27.1	25.2

[a] The climate variation is represented by: 1, annual mean temperature; 2, minimum temperature of the coolest month; 3, maximum temperature of the warmest month; 4, annual temperature range; 5, mean temperature of the coolest quarter; 6, mean temperature of the warmest quarter; 7, mean temperature of the wettest quarter; and 8, mean temperature of the driest quarter.

(Barnett and Lewis, 1978) are applied in reverse (i.e. to expand the influence of outliers rather than reduce their influence) to emphasize the effect of marginal records in each BIOCLIM climate attribute (Chapman, in prep.). Critical values are obtained for each specimen for each climate attribute. If the critical value is above the limit set for that sample size, the record is regarded as 'suspect' and is flagged for further checking. Tests have been successfully carried out on data sets varying in size from 4 to 476 records.

An example of this technique can be shown with *Eucalyptus robusta*. Specimen records of this species from the Queensland Herbarium are run through BIOCLIM to determine the climate profile (Table 11.1). The 100 records obtained are then graphed against various temperature attributes using a spread sheet package (Figure 11.4).

It can be seen from this analysis that three records are considerably out of step with the others for several temperature attributes. A similar pattern occurs if one uses the rainfall attributes. The plotted latitudes and longitudes for each of the records gives the results shown in Figure 11.5. The locality information supplied with the records shows that for one of the

suspect records (a), the descriptive locality given is 'Long Pocket CSIRO Labs'. The gazetteer (National Mapping, 1975) has four 'Long Pocket's in Queensland. Further examination of the locality information reveals that the correct latitude and longitude of the CSIRO Laboratories is different from that given. Another record (b) gives the descriptive locality as 'Kimburra', however no 'Kimburra' could be found in the gazetteer. The latitude and longitude given with the record is for 'Burra' and is obviously in error. Three other records (c) have latitude and longitude values that place them off shore. After these records are corrected and the data again processed using BIOCLIM, no further 'suspect' records are identified.

This technique is used extensively on data sets as they are loaded into the ERIN Specimen module. When suspect records are identified, a map (e.g. Figure 11.5) and a report for each species is returned to the custodian for checking and correction where necessary.

11.6 Development of a conservation strategy

Ecologically sustainable development (ESD), biodiversity, endangered species, comprehensive

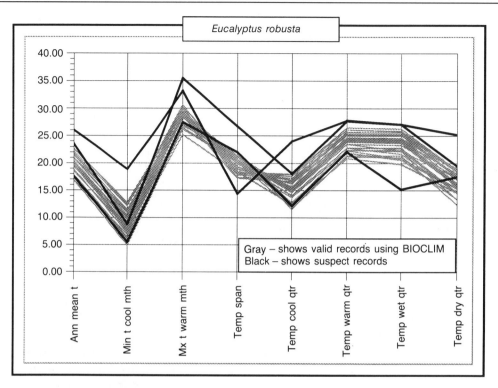

FIGURE 11.4 One hundred plant specimen records of *Eucalyptus robusta* plotted against eight temperature attributes (ann mean t = mean annual temperature; min t cool mth = mean minimum temperature of the coolest month; mx t warm mth = mean maximum temperature of the warmest month; temp span = mean temperature span; temp cool qtr = mean temperature of the coolest quarter; temp warm qtr = mean temperature of the warmest quarter; temp wet qtr = mean temperature of the wettest quarter; temp dry qtr = mean temperature of the driest quarter. The gray lines represent records that are not outliers and are regarded as valid records; the black lines represent those records that failed the outlier test and are regarded as 'suspect'.

reserve networks, natural heritage, state of environment reporting, and regional environmental management are all issues being hotly debated in a wide variety of Australia fora. A plethora of intersecting, often overlapping, and largely uncoordinated projects are currently under way or are being planned (Australian Academy of Science, 1992). One essential attribute that all these projects share is a requirement for comprehensive, reliable information about the Australian environment and its biota. The ERIN program is designed to build the necessary infrastructure to support these initiatives.

Two recent Australian Parliamentary inquiries

recommended that a bioregional framework be established across the continent for the planning and management of all environmental and natural resource programs (House of Representatives Standing Committee on Environment, Recreation and the Arts, 1992, 1993). The Committee stated that adoption of a bioregional framework is essential to the implementation and coordination of national ESD and biodiversity strategies. The purposes of a bioregional framework are:

- to develop a systematic basis for understanding and recognizing inherent biodiversity in each region

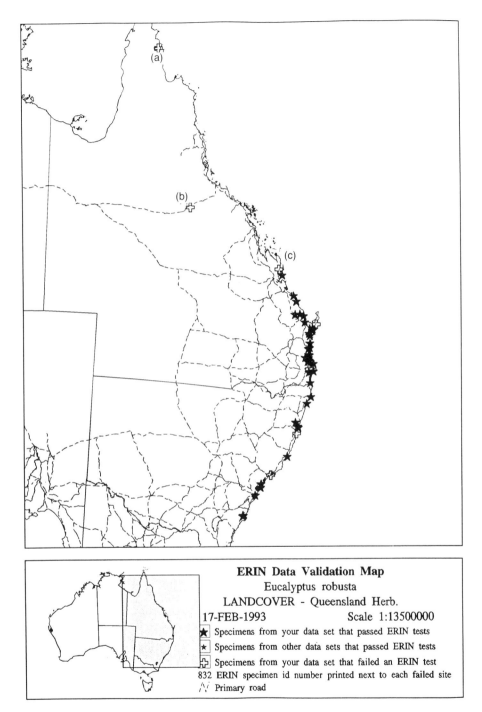

FIGURE 11.5 A map of 100 plant specimen records for *Eucalyptus robusta* in Queensland. Records marked 'a', 'b', and 'c' refer to the suspect records mentioned in the text. Data for this illustration were supplied by the Queensland Herbarium, Brisbane.

- to enable environmental auditing of each region to determine the conservation status of biodiversity threatening processes, sustainability of land use, and socio-economic issues so that conservation planning is focused and prioritized;
- to develop regional conservation strategies that integrate a representative reserve system with off-reserve measures and ecologically sustainable development.

ERIN has begun the development of bioregions for the continent, initially using attributes of soils, climate and topography as a surrogate for biodiversity information. ERIN has adopted the premise that, while maintaining flexibility in the collection of attributes and in classification methodologies, any regionalization must be credible, explicit and at an appropriate scale. Preliminary continental regionalizations based on 20, 30 and 40 regions are undergoing a process of assessment and validation (Thackway and Cresswell, 1992). The regionalizations are produced by classifying environmental objects into groups or types, based on the recorded association of environmental attributes for those objects. This process produces a mosaic of regions, usually presented as a map with a key to the groups or types present (Thackway, 1992). The regionalization shown in Figure 11.6 was derived at a resolution of 1/20th of a degree (ca. 5 km). Twelve attributes are consequently used to describe the temperature, precipitation, terrain and soils of 281 177 grid cells for the whole of Australia (Cresswell, 1992; Thackway and Cresswell, 1992). Classification of the data is carried out using a non-hierarchical method with nearest centroid sorting. This method has several advantages and is unaffected by the sequence of data units within the data set. It allows quite large data sets to be analyzed but requires considerable computing resources to do so (Malafant, 1992).

The preliminary regionalizations carried out by ERIN form the basis for discussions between conservation agencies. This focuses upon the bioregional approach, mentioned above, for use in environmental assessment and planning.

11.7 Long-term environmental monitoring

11.7.1 Introduction

The capacity to establish the present status of biodiversity is one thing. It is a different matter to monitor changes in biodiversity. Long-term environmental monitoring is essential, but few countries have made significant progress in this direction. Existing collections in association with paleohistorical records can give an indication of past changes for both geological and historical time periods. For example, in Australia, the dramatic range decline of a high proportion of the native mammal fauna can be documented from specimens collected over the past few centuries from areas where those species no longer occur.

The establishment of long-term ecological research sites is an essential element for environmental monitoring. A number of countries are attempting to address issues such as security of tenure and long-term financial support for environmental monitoring. Fortuitously, in a number of cases, countries have identified some long-running research sites and are beginning to build these into comprehensive national frameworks for long-term monitoring. Australia has yet to do the same.

11.7.2 Remote sensing

ERIN has identified remote sensing as a potentially valuable tool for monitoring environmental change. Security of supply of the imagery over the long term is an issue that is being addressed. ERIN is using 1 km resolution continental-scale imagery obtained from the NOAA AVHRR platform aggregated over a two-week period (to reduce the effects of

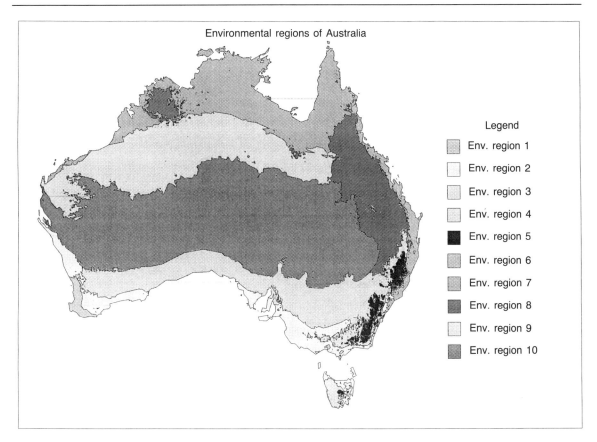

FIGURE 11.6 Ten group environmental regionalization of Australia using twelve environmental attributes of soil, climate and terrain and 281 177 grid cells. Thirty regions were derived using a non-hierarchical clustering technique. The 30 groups were then aggregated into 10 super groups using environmental relationships between each of the regions.

cloud). This is being used to monitor changes in landcover as well as to monitor fires, floods, and large water bodies such as wetlands (Bullen, 1992). Phenological patterns are evaluated over a series of 12-monthly periods to detect long-term changes in vegetation cover. Changes over shorter periods are used to monitor fire in areas of northern Australia where large areas of the country are burnt each year.

11.8 Summary

Australia is developing a continental environmental spatial information system, through the Environmental Resources Information Network (ERIN), to underpin the development and implementation of environmental policy. This system is being designed around key patterns and processes in Australian landscapes. A key project is the coordinated development of data sets on those plant species that dominate Australian vegetation. Collaborative efforts to integrate collections of plant information from around the nation, along with development of a national plant list, are giving Australia the capacity to analyze and monitor its biodiversity. Spatial information systems are used to link these and other data sets with bioclimatic modeling and remote sensing to provide an integrated framework for environmental decision making.

Acknowledgements

Many of the ideas in this paper are the result of work carried out by a number of ERIN Unit staff. In particular, we would like to thank Frank Bullen and Kate Sanford-Readhead for support in preparing the illustrations and Ann Bull, Ian Cresswell and Stewart Noble for comments on the manuscript.

References

ANZECC (1991) *List of Endangered and Vulnerable Vertebrate Fauna, April 1991*, Australian National Parks and Wildlife Service, Canberra.

Armstrong, J.A. (1992) The funding base for Australian biological collections. *Australian Biologist*, 5(1), 80–8.

Australian Academy of Science (1992) *Global Change. A Research Strategy for Australia 1992–1996*, Australian Academy of Science, Canberra.

Barnett, V. and Lewis, T. (1978) *Outliers in Statistical Data*, Wiley, Chichester, UK.

Bennett, S., Brereton, R., Mansergh, I. *et al.* (1991) *The Potential Effect of the Enhanced Greenhouse Climate Change on Selected Victorian Fauna*. Arthur Rylah Institute Tech. Rep. No. 123, Department of Conservation and Environment, Victoria.

Bolton, M.P. (ed.) (1992) *Vegetation. From Mapping to Decision Support*. A workshop to establish a set of core attributes for vegetation. Environmental Resources Information Network, Canberra.

Booth, T.H. (1985) A new method for assisting species selection. *Commonwealth Forestry Review*, 64, 241–50.

Booth, T.H., Nix, H.A., Hutchinson, M.F. and Busby, J.R. (1987) Grid matching: a new method for homocline analysis. *Agriculture and Forestry Meteorology*, 39, 241–55.

Bridgewater, P. (1991) Impacts of climate change on protected area management. *Environment Professional*, 13, 74–78.

Bullen, F.B. (1992) ERIN's use of NOAA AVHRR data for environmental monitoring. *Erinyes*, 14, 8–9.

Busby, J.R. (1986a) A biogeographic analysis of *Nothofagus cunninghamii* (Hook.) Oerst. in southeastern Australia. *Australian Journal of Ecology*, 11, 1–7.

Busby, J.R. (1986b) *Bioclimatic Prediction System (BIOCLIM). User's Manual Version 2.0*, Australian Biological Resources Study, Canberra.

Busby, J.R. (1988) Potential impacts of climate change on Australia's flora and fauna, in *Greenhouse: Planning for Climate Change* (ed. G.I. Pearman), CSIRO, Melbourne, pp. 387–98.

Busby, J.R. (1991) BIOCLIM – A bioclimatic analysis and prediction system, in *Nature Conservation: Cost Effective Biological Surveys and Data Analysis* (ed. C.R. Margules and M.P. Austin), CSIRO, Canberra, pp. 64–8.

Chapman, A.D. (1991a) Land cover project, *Erinyes*, 9, 4–6.

Chapman, A.D. (1991b) *Australian Plant Name Index A–C*, Australian Flora and Fauna Series No. 12, pp. 1–898, Australian Government Publishing Service, Canberra.

Chapman, A.D. (1991c) *Australian Plant Name Index D–J*, Australian Flora and Fauna Series No. 13, pp. 899–1710, Australian Government Publishing Service, Canberra.

Chapman, A.D. (1991d) *Australian Plant Name Index K–P*, Australian Flora and Fauna Series No. 14, pp. 1711–2476, Australian Government Publishing Service, Canberra.

Chapman, A.D. (1991e) *Australian Plant Name Index Q–Z*, Australian Flora and Fauna Series No. 15, pp. 2477–3055, Australian Government Publishing Service, Canberra.

Chapman, A.D. (1992) Quality control and validation of environmental resource data, in *Data Quality and Standards: Proceedings of a Seminar Organised by the Commonwealth Land Information Forum, Canberra, 5 December 1991*, Commonwealth Land Information Forum, Canberra.

Chapman, A.D. (in prep.) The use of climate for detecting outliers in geo-referenced biological records.

Cogger, H.G. (1983) Amphibia and reptilia, in *Zoological Catalogue of Australia* (ed. D.W. Walton), Vol. 1, Australian Government Publishing Service, Canberra.

Cresswell, I. (1992) Environmental regionalisations for the Environmental Resources Information Network, in *Environmental Regionalisation. Establishing a Systematic Basis for National and Regional Conservation Assessment and Planning* (ed. R. Thackway), Environmental Resources Information Network, Canberra, pp. 66–8.

Croft, J.R. (1988) *Herbarium Information Standards and Protocols for Interchange of Data*, Australian National Botanic Gardens, Canberra.

Gibson, N. (1986) *Huon Pine Conservation and Management*. Wildlife Division Technical Report 86/3, National Parks and Wildlife Service, Tasmania.

Hawke, R.J.L. (1989) *Our Country Our Future*. Statement on the Environment, Australian Government Publishing Service, Canberra.

Hill, R.S., Read, J. and Busby, J.R. (1988) The temperature-dependence of photosynthesis of some Australian temperate rainforest trees and its biogeographical significance. *Journal of Biogeography*, **15**, 431–49.

Hnatiuk, R.J. (1990) *Census of Australian Vascular Plants*, Australian Flora and Fauna Series No. 11, pp. 1–650, Australian Government Publishing Service, Canberra.

House of Representatives Standing Committee on Environment, Recreation and the Arts (1992) *Biodiversity. The Contribution of Community-Based Programs*, Australian Government Publishing Service, Canberra.

House of Representatives Standing Committee on Environment, Recreation and the Arts (1993) *Biodiversity. The Role of Protected Areas*, Australian Government Publishing Service, Canberra.

Leigh, J.H. and Briggs, J.D. (1992) *Threatened Australian Plants*, Australian National Parks and Wildlife Service, Canberra.

Lindenmeyer, D.B., Nix, H.A., McMahon, J.P. and Hutchinson, M.F. (1990) Bioclimatic modelling and wildlife conservation and management: A case study on Leadbeater's Possum, *Gymnobelideus leadbeateri*, in *Management and conservation of Small Populations* (ed. T.W. Clark and J.H. Seebeck), Chicago Zoological Society, Illinois, USA.

Lindenmeyer, D.B., Nix, H.A., McMahon, J.P. *et al.* (1991) The conservation of Leadbeater's possum, *Gymnobelideus leadbeateri* (McCoy): A case study of the use of bioclimatic modelling. *Journal of Biogeography*, **18**, 371–83.

Longmore, R.C. (1986) *Atlas of Australian Elapid Snakes*, Australian Flora and Fauna Series 7, pp. 1–115, Australian Government Publishing Service, Canberra.

McKenzie, G.M. and Busby, J.R. (1992) A quantitative estimate of Holocene climate using a bioclimatic profile of *Nothofagus cunninghamii* (Hook.) Oerst., *Journal of Biogeography*, **19**, 531–40.

Malafant, K. (1992) Murray Darling Basin ecosystem analysis, in *Environmental Regionalisation. Establishing a Systematic Basis for National and Regional Conservation Assessment and Planning* (ed. R. Thackway). Environmental Resources Information Network, Canberra, pp. 62–5.

Markgraf, V., Bradbury, J.P. and Busby, J.R. (1986) Paleoclimates in southwestern Tasmania during the last 13,000 years. *Palaios*, **1**, 368–80.

Miles, H.M. (1992) *Review of the Environmental Resources Information Network [ERIN]*, Australian National Parks and Wildlife Service, Canberra.

National Mapping (1975) *Australia 1:250,000 Map Series Gazetteer*, Australian Government Publishing Service, Canberra.

Nobs, R.F. (1990) Indicators for monitoring biodiversity: A hierarchical approach. *Conservation Biology*, **4**, 355–64.

Panetta, F.D. and Dodd, J. (1987) Bioclimatic prediction of the potential distribution of skeleton weed *Chondrilla juncea* L. in Western Australia. *Journal of the Australian Institute of Agriculture Science*, **53**, 11–16.

Prendergast, H.D.V. and Hattersley, P.W. (1985) Distribution and cytology of Australian Neurachne and its allies (Poaceae), a group containing C3, C4 and C3–C4 intermediate species, *Australian Journal of Botany*, **33**, 317–36.

Richardson, B.J. and McKenzie, A.M. (1991) *Australia's Taxonomists and Taxonomic Collections*, Australian National Parks and Wildlife Service, Canberra.

Thackway, R. (ed.) (1992) *Environmental Regionalisation. Establishing a Systematic Basis for National and Regional Conservation Assessment and Planning*, Environmental Resources Information Network, Canberra.

Thackway, R. and Cresswell, I.D. (1992) *Environmental Regionalisations of Australia: A User Oriented Approach*, Environmental Resources Information Network, Canberra.

Part Eight

Possibilities for the Future

Chapter twelve

Possibilities for the future

Ronald I. Miller

12.1 Introduction

To conserve nature in today's world, the collective impact of protecting similar natural units across diverse scales of space and time needs to be considered. In other words, to ensure the conservation of a single type of an ecosystem or habitat or species within a single region or country, all the natural units of this type need to be considered collectively. The influence of a change in scale in either time or space always needs to be weighed to ensure that the appropriate natural units can be maintained for the foreseeable future.

For a single species, the consideration of predictable environmental variation across space (e.g. succession or human development) and time (e.g. climatic or hydrologic changes) is critical for long-term preservation. The impact of these variation patterns on the collective natural units of the species can effectively be modeled and represented using maps. The forefront of future map utilization in conservation planning will be the presentation of the effects of spatial and temporal scale changes on collections of similar natural units.

For scientific accuracy, the representation of the spatial variation of the elements of nature on a map requires a careful selection process. The factors regulating the spatial variation of these natural elements require their presen-tation at an appropriate scale (Jensen, 1992). Therefore only factors with compatible appropriate scales should be presented on the same map. The characteristic scales and frequencies of ecosystem processes will only be identified with the accumulation of knowledge gained from additional ecological research.

12.2 Applications of GIS

Since the development of GIS technology, very practical factors have limited the use of GIS by conservationists. For example, problems related to tracking and understanding the impact of the data-processing steps that lead to output products from the GIS (Davis *et al.*, 1991) will need to be more clearly delineated in the future. Other problems include the required training and expertise to operate the technology effectively and efficiently, the development of practical applications for down-to-earth conservation problems, the time required for database development, and the cost of software, hardware and expertise. This book presents many successful, practical applications of computerized mapping that clearly demonstrate the current day usefulness of computer technology to conservation in the twentieth century and beyond.

Our perceptual limits, and the logistical

Mapping the Diversity of Nature. Edited by Ronald I. Miller.
Published in 1994 by Chapman & Hall, London. ISBN 0 412 45510 2.

restrictions imposed by maps, constrain the scales at which we can consider spatial distribution patterns of natural features. Limited field data also constrain our ability to represent the detail of natural features on the small map scale. Data limitations therefore currently constrain our ability to capitalize on all of the facilities of GIS. In the near future, the accurate depiction of species and habitat distribution patterns will require further research on the influence of data error and uncertainty.

In the future, it is necessary that we define the appropriate strategies for spatial modeling and data acquisition based upon known conditions in the field. For maps at the regional and global scales, this will require research on the scale dependence of surface features and on the absolute versus relative scales within the remote-sensing environment.

For regional and global assessments, the influence of spatial variability and of spatial, spectral and temporal resolution need to be identified. We must also identify the appropriate sampling and scaling strategies for use with sparse ground measurements. The current research issues in this field are many (e.g. Davis *et al.*, 1991; Clark *et al.*, 1991).

Effective management of natural resources requires knowledge about the scales at which human impacts should be monitored and examined (Hunsaker *et al.*, 1993). In the past, ecological modeling focused upon state changes at one location over time (ibid.). The introduction of sophisticated GIS capabilities allows ecologists to consider species changes in space and time within much more dynamic environmental contexts (e.g. Johnson and Worobec, 1988). For example, a recent model of fire hazard detection uses a GIS as a land management decision support system (Kessell, 1990). This same general modeling approach certainly will be used in the future for managing the linkages between conservation and development.

12.3 Future applications in ecology

Biology is set apart from other data-rich fields by its complex and heterogeneous data environment rather than from either the sheer volume of data produced by ecological surveys (Kingsbury, 1993) or the volume of data required for analysis. This complicated data environment has so far limited the usefulness of satellite imagery data in ecology. The future valid application of satellite imagery data in ecology will require many refinements that will include (Wessman, 1992):

(1) stratified sampling procedures to address biospheric processes that vary in time and space
(2) regional extrapolations that identify controlling factors
(3) the use of gradients that can be remotely sensed from space
(4) time series of remotely sensed data (necessary for studies of global change)
(5) a hierarchy of spatial resolution measurements to address mesoscale heterogeneity for generalizations to global scales
(6) global geographic databases that represent parameters such as soils, vegetation, land use, etc.
(7) the integration of ecological models at the process level for the extrapolation of local to regional measurements.

The measurements needed for effectual modeling of ecological systems at the regional and global scales may become available in the near future (Wessman, 1992; Hunsaker *et al.*, 1993). However, for the effective long-term management and protection of natural resources, a better understanding of how the scale of an environmental hazard affects ecological processes will be essential (Hunsaker *et al.*, 1993).

The true integration of the spatial dimension into ecological modeling will only occur when ecologists can define the natural units of landscapes and how the state of one unit affects the future state of another. Most importantly, ecologists will always need to test at regional and global scales whether or not the future state of a landscape unit is independent of adjacent units. For modern-day conservation and development, the level of resolution included in a model must fit the resolution required by the problem at hand. Modelers must learn to compromise between the known complexity of ecosystems and the need to address a problem with limited data and within a limited time period (Grant, 1988).

12.4 Some promising future mapping approaches for conservation

Ecological patterns and processes comprise the fundamental components of biodiversity in the natural world (Perry, 1993). Natural patterns, both in time and space, can often be well represented on maps. The production of maps and the representation of the distribution patterns of natural features has advanced in recent years with the development of the GIS. The GIS has now successfully passed through its infancy stage. GIS tools and approaches are beginning to emerge that will allow scientists to perform a more accurate analysis of ecological patterns and processes and to provide some causal underpinning for many spatial patterns in nature. This greater understanding will enhance the endurance of implemented biodiversity conservation practices and programs. Some of the most recent technological tools becoming available for conservation planning are presented below.

12.4.1 Videography

A critical gap remains between satellite data and the many varieties of field observations. As stated above, our inability to verify broad-scale mapping efforts using field survey data persists as a significant obstacle in our attempts to map natural features. A new approach that combines GPS and videography offers a practical methodology to bridge this gap (Graham, 1993; Hassan and Hutchinson, 1992, p. 11). This method comprises a set of procedures for data acquisition, interpretation and verification that are based upon integrated airborne GPS/video systems. Currently this method is sucessfully being used to validate and classify TM imagery to produce vegetation maps for the state of Arizona (Graham, 1993). This new approach may provide a method to obtain extensive and inexpensive information within a short time frame. This will permit the verification and validation of natural features required for the production of accurate maps at relatively small map scales.

12.4.2 AVIRIS: A new imagery technology

The capability to distinguish the identities of vegetation elements from imagery data is a function of the interpretation of emission and absorption features of the light spectrum (Clark *et al.*, 1992). All the applications of satellite data presented in this book result from imagery data derived from the infrared portion of the light spectrum. The AVHRR, MSS, TM and SPOT technologies are all based upon images recorded within the infrared portion of this spectrum. The greater wavelength of the infrared versus the visual portion of the light spectrum permits less focus of photographic images. In addition, though infrared data may be defined as having a high spatial resolution per pixel (e.g. 30 m^2 in the case of the TM technology), the data still need to be integrated across enough pixels to obtain adequate resolution.

A new technology, AVIRIS (which is currently an aircraft flown technology), records imagery data across the entire light spectrum and provides the potential for a significantly higher resolution of imagery data for the production of landcover and vegetation maps

(Green, 1990, 1991, 1992). Global coverage using AVIRIS will become available in the near future when AVIRIS becomes operational on a satellite.

To date, AVIRIS data have primarily been used for making direct measurements of canopy chemistry (e.g. protein, lignin and nitrogen – Roberts *et al.*, 1992). However, AVIRIS research shows great promise for high resolution detailed vegetation mapping produced directly from imagery spectrometer data. A new approach to the interpretation of these data is now being explored that provides detailed identification of ecological elements on the earth's surface (ibid.). In this approach, image classes are allowed to vary from pixel to pixel, and this provides a significantly greater capability for identifying landcover detail. In the future, the AVIRIS data may yield very fine-scale resolution data for the identification of many plant species across the landscape. This will significantly enhance the detail and usefulness of vegetation map data derived from satellite imagery.

12.4.3 Grid mapping

Species distribution mapping is an increasingly important part of ecological science world wide (e.g. Rotshil'd and Fedotov, 1988; Stevens and Goodson, 1993). The equal-area grid arrangement is a useful framework for representing species presence/absence data and for analyzing species distribution patterns. This approach is particularly cogent in developing parts of the world where the primary sources of information about species spatial distributions are very often presence/absence data. The appropriateness of this approach is demonstrated in several recent studies of birds in East Africa (Miller *et al.*, 1989; Pomeroy, 1989, 1993). Many recent SSC publications also use the grid approach to monitor threatened species in many parts of the world. In Chapter 10, a refined equal-area procedure was introduced

that is producing a global coverage of coral reef fish species for conservation management. This method appears to permit monitoring on the regional and global scales and it will certainly be used more extensively in the future as an approach to conservation in the international arena.

12.5 Future uses of spatial data in conservation

Today the monitoring and conservation of nature requires information that is registered to geographic locations. This is exemplified well by the modern conservation monitoring techniques used in the UK, a country with a rich history of dedicated environmental vigilance (Goldsmith, 1991). Spatial data can serve both as a catalyst and an inhibitor in environmental modeling, and can bring together information from different sources as catalysts for environmental research and planning. At the same time, spatial data from many disparate sources create some difficulties, as pointed out in the chapters of this book.

Rapid assessment strategies are now being used by several notable international conservation organizations (Abate, 1992; Hassan and Hutchinson, 1992, p. 110). This strategy involves deploying small teams of biological experts to a region of the world in need of conservation attention. These experts then collaborate to produce maps that identify the habitats and species in need of immediate protection. Recently, Conservation International has successfully used this approach in Manaus, Brazil (1990) and in Papua New Guinea (1992).

The establishment of a National Biological Survey in the United States will be a positive advancement for the management and conservation of US biological resources. Yoon (1993) has stated that this agency will 'probably focus upon acquiring and consolidating information about US ecosystems and their constituents', and the mapping and database approaches presented in this book will serve as some of the basic approaches to be used by this new effort in the US.

12.6 Mapping and modeling at the global scale

As mentioned previously, spatial scale is a particularly difficult problem when global phenomena are modeled. One reason for this difficulty is the need to extrapolate from plot data at the large map scale to the small map scale for the production of maps useful to agencies with broad regional constituencies. This extrapolation increases the uncertainty of the data depicted in the regional models. The creation of valid models and maps to depict the possible scenarios for transformations predicted to be caused by global change is inextricably linked to the issues of scale (Wessman, 1992).

The majority of so-called map data are generated by contouring methods that are based upon a finite number of observations. Propagation of uncertainties from these data pose significant problems for spatial models (Hunsaker *et al.*, 1993). One solution to the problem of moving data between map scales is the development of models that are unique at each scale. Some of the leading applications in this area are being developed at Colorado State University. These are called mesoscale models and each model uses a different classification scheme for each unique scale (Evans and Alberti, 1993).

12.7 Conservation and development

The Environment Department at the World Bank is currently undertaking a series of mapping projects in an effort to develop approaches that integrate the needs of biodiversity protection, protected areas and indigenous peoples. These projects are focused upon three countries

that have World Bank funded biodiversity protection projects and large and diverse groups of indigenous peoples. These projects are particularly interested in collecting data about the participation of indigenous and other local communities in protected area design and management. Biodiversity maps will be produced in these projects along with public policy papers. As similar projects are introduced into developing countries in future years, many of the the approaches to biodiversity mapping presented in these chapters will play a prominent role.

Computer hardware and software are currently chasing each other into the future. But a large part of human civilization is being left out of the contemporary conceptual framework created by computer technology. Successful projects in developing countries need to concentrate on helping people to practically improve their lives in ways that coincidentally protect the nature around them. In the past, projects for biodiversity protection have often become highly focused upon computer technology. These projects need to focus more on who will use the GIS and the associated databases, how they will make daily use of these tools, and how these tools will be integrated into management decision making. Computers need to be viewed only as practical tools that can help the people in these countries to address their problems. The most effective way that computers can be used to expedite the integration of conservation and development in the future is with the use of simple and practical mapping and database methods.

Both conservation and development organizations can use the mapping and database techniques presented here to produce practical products that improve the efficacy of conservation in the world today. The techniques presented in these pages include methods for the in-country monitoring of biodiversity, the development of planning capabilities to support both national and continent-wide conservation strategies, and the identification of key hotspots (Prendergast *et al.*, 1993) of high biological diversity for proactive planning. This book provides a state-of-the-art review of these rapidly developing techniques.

References

Abate, T. (1992) Environmental rapid assessment programs have appeal and critics. *Bioscience*, **42**(7), 486–9.

Clark, D.M., Hastings, D.A. and Kineman, J.J. (1991) Global databases and their implications for GIS, in *Geographical Information Systems*, Vol. 2: *Applications* (ed. D.J. Maguire, M.F. Goodchild and D.W. Rhind), Longman, Harlow, Essex, UK.

Clark, R.N., Swayze, G.A., Koch, C. and Ager, C. (1992) Mapping vegetation types with the multiple spectral feature mapping algorithm in both emission and absorption. *Summaries of the Third Annual JPL Airborne Geoscience Workshop*, June 1–5, 1992, Vol. 1, *AVIRIS Workshop* (ed. Robert O. Green), Pasadena, California, pp. 60–2.

Davis, F.W., Quattrochi, D.A., Ridd, M.K. *et al.* (1991) Environmental analysis using integrated GIS and remotely sensed data: Some research needs and priorities. *Photogrammetric Engineering and Remote Sensing*, **57**, 689–97.

Evans, J. and Alberti, M. (1993) Sharing spatial information systems across environmental agencies, regions, and scale: Comparing theory with experiences. Presented at: *Environmental Information Theory and Analysis: Ecosystem to Global Scales*. An international symposium sponsored by the US National Science Foundation, 20–22 May, 1993, University of New Mexico, Albuquerque, New Mexico.

Goldsmith, F.B. (ed.) (1991) *Monitoring for Conservation and Ecology*, Chapman & Hall, London.

Graham, L. (1993) Airborne video for near real-time natural resource applications. *Journal of Forestry*, **91**, 28–32.

Grant, W.E. (1988) Models for conservation and wildlife management. *Ecological Modeling*, **41**, 325–6.

Green, R.O. (ed.) (1990) *Proceedings of the Second Airborne Visible/Infrared Imaging Spectrometer (AVIRIS) Workshop*, JPL 90–54. Jet Propulsion Laboratory, California Institute of Technology, Pasadena, California.

Green, R.O. (ed.) (1991) *Proceedings of the Third Airborne Visible/Infrared Imaging Spectrometer (AVIRIS) Workshop*, JPL 91–28. Jet Propulsion Laboratory, California Institute of Technology, Pasadena, California.

Green, R.O. (ed.) (1992) *Summaries of the Third Annual JPL Airborne Geoscience Workshop*, JPL 92–14, Vol. 1. Jet Propulsion Laboratory, California Institute of Technology, Pasadena, California.

Hassan, H.H., and Hutchinson, C. (eds) (1992) *Natural Resource and Environmental Information for Decisionmaking*. The World Bank, Washington, D.C.

Hunsaker, C.T., Nisbet, R.A., Lam, D. *et al.* (1993) Spatial models of ecological systems and processes: The role of GIS, in *Environmental Modeling with GIS* (ed. M. Goodchild, B. Parks and L. Steyaert), Oxford University Press, New York, pp.248–64.

Jensen, M. (1992) Ecological system description and mapping, in *Stewardship First: Taking an Ecological Approach to Management*, Proceedings of a USDA National Workshop, April 27–30, 1992, Salt Lake City, Utah, pp.91–108.

Johnson, D.L. and Worobec, A. (1988) Spatial and temporal computer analysis of insects and weather: Grasshoppers and rainfall in Alberta (Canada). *Memoirs of the Entomological Society of Canada*, **146**, 33–48.

Kessell, S.R. (1990) An Australian geographical information and modelling system for natural area management. *International Journal of Geographical Information Systems*, **4**(3), 333–62.

Kingsbury, D.T. (1993). *Research Opportunities in Computational Biology: Results from a series of Invitational Workshops*. Draft report produced for the US National Science Foundation.

Miller, R.I., Stuart, S.N. and Howell, K.N. (1989). A methodology for analyzing rare species distribution patterns utilizing GIS technology: The rare birds of Tanzania. *The Journal of Landscape Ecology*, **2**(3), 173–89.

Perry, D.A. (1993) Biodiversity and wildlife are not synonomous. *Conservation Biology*, **7**(1), 204–5.

Pomeroy, D. (1989) Using East African bird atlas data for ecological studies. *Annals of the Zoological Fennici*, **26**, 309–14.

Pomeroy, D. (1993) Centers of high biodiversity in Africa. *Conservation Biology*, **7**(4), 901–7.

Prendergast, J.R., Quinn, R.M., Lawton, J.H. *et al.* (1993). Rare species, the coincidence of diversity hotspots and conservation strategies. *Nature*, **365**, 335–7.

Roberts, D.A., Smith, M.O., Sabol, D.E. *et al.* (1992) Mapping the spectral variability in photosynthetic and non-photosynthetic vegetation, soils, and shade using AVIRIS. *Summaries of the Third Annual JPL Airborne Geoscience Workshop*, June 1–5, 1992, Vol 1, *AVIRIS Workshop* (ed. Robert O. Green), Pasadena, California, pp. 38–40.

Rotshil'd, E.E. and Fedotov, V.G. (1988) The experience mapping mammal population with ecological data. *Byulleten' Moskovskogo Obshchestva Ispytatelei Prirody Otdel Biologicheskii*, **93** (5), 114–26.

Stevens, D.R., and Goodson, N.J. (1993) Assessing effects of removals for transplanting on a high-elevation Bighorn sheep population. *Conservation Biology*, **7** (4), 908–15.

Wessman, C.A. (1992) Spatial scales and global change: Bridging the gap from plots to GCM grid cells, in *Annual Review of Ecology and Systematics* (ed. D.G. Fautin, D.J. Futuyma and F.C. James). Vol. 23. Annual Reviews Inc., Palo Alto, California.

Yoon, C.K. (1993) Counting creatures great and small. *Science*, **260**, 620–2.

Glossary

Association table A data matrix that associates individual species with vegetation types, counties, or other mapped features. Columns represent species and rows represent the map feature. If species j is associated with a particular feature, k, a 1 is coded into cell k, j.

Biodiversity The living elements that constitute the natural world – as distinct from the man-made elements of our world.

Biodiversity hotspots Those regions that have notably higher numbers of species of a particular taxonomic group than the surrounding areas.

California Natural Diversity Database This database is maintained by the Natural Heritage Division of the California Department of Fish and Game, consisting of observations of sensitive wildlife species and communities in the state.

California Wildlife Habitat Relationships Database This database consists of information on the habitat preference of 646 species of terrestrial vertebrates resident in California.

Cell resolution Represents the grain of the data represented by each cell.

Data extent The total span of data under consideration, either in terms of spatial area or duration of time.

Ecochores Landscape compartments represented by ecotope assemblages with common features.

Ecotope Spatial representation of an ecosystem on a site (Naveh, 1984). Ecotope types are often represented by vegetation units and are used for characterizing the landscape.

Endemic Bird Area An area that includes the entire distribution of at least two restricted-range bird species.

Extent Area of analysis indicated by a specified set of results.

GIS additive overlay procedure A procedure that overlays and combines overlay procedure multiple map layers into a single map layer. The number of coincident occurrences in the overlain map layers is accumulated into the new map layer.

Ground truth Data collected from field surveys or earth-based instruments and measurements.

Habitat feature data These are confirmed, field collected, mapped data that represent habitat features.

Habitat availability Habitat availability for a species can be defined by the geographical distance between habitat types that are known to contain the species from previous observations.

Inter-point interval The distance between each digitized point on a map.

Large map scale Comparatively little reduction is necessary for the map representation. A map at the 'large' scale contains great amounts of information but it covers only a small area on the ground. This is a relative term for a map with a reduction ratio of 1:50 000 or less (e.g. 1:25 000) (Robinson *et al.*, 1984).

Map scale The ratio or proportion between the map dimensions and the 'ground' dimensions.

Megadiversity country A country characterized by high levels of biodiversity relative to other countries (e.g. Brazil, Mexico and Madagascar).

Mesoscale models Models in which each scale analyzed employs a unique classification scheme.

Massenerhebung The retention of heat by large masses of high terrain. Small masses of terrain of equivalent altitude and latitude would heat up and cool down more rapidly. Where the phenomenon is found, lowland vegetation reaches higher elevations than are typical.

Mist-net sampling grid The methodology used in Chapter 6 that comprised 13 mist-nets (12 m × 2.6 m × 36 mm mesh) placed in five lines over a 1 hectare plot.

Multiple Species Conservation Plan A conservation plan for western San Diego County, USA.

Natural resources data Mapped digital data are now available for many areas in the world. These are environmental data collected using a remote-sensing technique. These data may or may not have been field checked and they usually represent climatic, geologic and vegetation patterns.

Natural unit A mapped unit of nature that is considered to be homogeneous for some character. These are cartographically considered to be homogeneous units to expedite their identification on a map. Scientifically, these units can all be further divided. Examples of natural units: (1) a mapped area, bounded by an elevation contour, which is defined to be within a single elevation range; (2) a distribution record for a single species; (3) a polygon which represents a single occurrence of a habitat or ecosystem type.

Normalized Difference Vegetation Index NDVI = (near infrared) − (visible red)/(near infrared) + (visible red)

Provenance The native origin of a tree variety; the place of origin.

Pseudo-endemic Species whose known range is geographically restricted and significantly smaller than its actual range. This causes it to appear to be endemic in the region of its known occurrence.

Resource sectors A collective term for the institutions and personnel concerned with the development and management of renewable resources (i.e. agricultural, forestry, fisheries, wildlife, and conservation expertise).

Scale The interval of space or time over which a measurement is made (Davis *et al.*, 1991).

Small map scale A comparatively great reduction is necessary for map representation. A map at the 'small' scale contains a very small amount of detailed information but it covers a very large area on the ground. This is a relative term for a map with a reduction ratio of 1:50 000 or more (e.g. 1:1 000 000) (Robinson *et al.*, 1984).

Smooth cell count The sum of the number of species that occur in a single cell and adjacent cells whose centers are within three degrees of the cell being smoothed.

SPOT The *Système pour l'Observation de la Terre* (SPOT) satellite carries two high resolution visible (HRV) pushbroom sensors. One sensor operates in multispectral mode and collects three channels: a green channel, a red channel, and a near-infrared channel. This multispectral sensor has a spatial resolution of 20 meters. The sensor in panchromatic mode collects a single image with a spatial resolution of 10 meters.

Species distribution Defines the known locations of a species within a designated region.

Species range The spatial area across which a species may be found. For example, the range of a migratory bird species would include areas along the migration route as well as areas used for longer term habitation. This term refers more to predicted patterns of species occurrence than does the **species distribution**.

Spot distribution map A map with species' geographic ranges presented as spots or other symbols that mark the sites of known occurrences. This contrasts with a map that shows the ranges with lines around the periphery of the range. This approach may conceal gaps inside the periphery of the range.

Taxonomy The science of classifying and understanding the structural differentiation that characterizes life.

Variable circular A method of censusing birds, developed by Reynolds *et al.* (1980). The number of individuals and the distance to each individual (up to 50 m from the observer) seen or heard within a 5-minute time period are recorded for each census point.

References

Davis, F.W., Quattrochi, D.A., Ridd, M.K. *et al.* (1991) Environmental analysis using integrated GIS and remotely sensed data: Some research needs and priorities. *Photogrammetric Engineering and Remote Sensing*, 57, 689–97.

Naveh, Z. (1984) Conceptual and theoretical basis of landscape ecology as a human ecosystem science, in *Landscape Ecology: Theory and Application* (ed. Z. Naveh and A.S. Lieberman), Springer-Verlag, New York.

Reynolds, R.T., Scott, J.M. and Nussbaum, C.A. (1980) A variable circular-plot method for estimating bird numbers. *Condor*, 82, 309–13.

Robinson, A.H., Sale, R.D., Morrison, J.L. and Muehrcke, P.C. (1984) *Elements of Cartography* (5th edn), Wiley, New York, 544pp.

List of Acronyms

AED	African Elephant Database
AESG	African Elephant Specialist Group
ANPP	Aerial Net Primary Productivity
AVHRR	Advanced Very High Resolution Radiometer
AVIRIS	Airborne Visible/Infrared Imaging Spectrometer
BIOCLIM	Bioclimate Analysis and Prediction System
CAR	Central African Republic
CITES	Convention on International Trade in Endangered Species
CNDDB	California Natural Diversity Database
DBMS	Database Management System
EBA	Endemic Bird Area
EIR	Environmental Impact Report
ERIN	Environmental Resources Information Network
ESD	Ecologically Sustainable Development
ESRI	Environmental Systems Research Institute
FAO	Food and Agriculture Organization of the UN
GEMS	Global Environment Monitoring System
GIS	Geographic Information System
GPS	Global Positioning System
GRID	Global Resource Information Database
HRV/MLA	High Resolution Visible/ Multispectral Linear Array
ICBP	International Council for Bird Preservation, now designated BirdLlife International
IUCN	The World Conservation Union
IGIS	Integrated Geographic Information System
KWS	Kenya Wildlife Service
LEP	Laikipia Elephant Project
MSCP	Multiple Species Conservation Plan
MSS	Multispectral Scanner
NDVI	Normalized Difference Vegetation Index
NGO	Non-Governmental Organization
NOAA	National Oceanographic and Atmospheric Administration
ONC	Operational Navigational Charts
PC	Personal Computer
RAP	Rapid Assessment Procedures
RNU	Regional Natural Units
SPOT	*Système pour l'Observation de la Terre*
SSC	IUCN Species Survival Commission
SIS	Spatial Information System

TM	Thematic Mapper	WCI	Wildlife Conservation International
TWINSPAN	Two-way Indicator Species Analysis	WCMC	World Conservation Monitoring Centre
UNEP	United Nations Environment Programme	WHR	California Wildlife Habitat Relationships database
UTM	Universal Transverse Mercator		
VCP	Variable Circular Plot	WWF	World Wide Fund for Nature

Index

Page numbers in **bold** type refer to figures, and those in *italic* refer to tables.